T0275979

SpringerBriefs in Applied Sciences and Technology

PoliMI SpringerBriefs

More information about this series at http://www.springer.com/series/11159
http://www.polimi.it

Simone Ferrari · Valentina Zanotto

Building Energy Performance Assessment in Southern Europe

Simone Ferrari
Politecnico di Milano
Milan
Italy

Valentina Zanotto
Amstein + Walthert AG
Zürich
Switzerland

ISSN 2191-530X ISSN 2191-5318 (electronic)
SpringerBriefs in Applied Sciences and Technology
ISSN 2282-2577 ISSN 2282-2585 (electronic)
PoliMI SpringerBriefs
ISBN 978-3-319-24134-0 ISBN 978-3-319-24136-4 (eBook)
DOI 10.1007/978-3-319-24136-4

Library of Congress Control Number: 2015950018

Springer Cham Heidelberg New York Dordrecht London

Printed on acid-free paper

Springer International Publishing AG Switzerland is part of Springer Science+Business Media
(www.springer.com)

Preface

Due to the prevailing climatic conditions, buildings in most European countries are characterised by a greater need for heating than cooling.

Several simplifications can be generally adopted for the calculation of the thermal balance of a building when considering the heating side. In fact, the assessment of the heating energy needs of a building mainly depends on the amount of indoor air volumes to be renewed to ensure the well-being of the occupants, regardless of the construction features, and on the performance of the building envelope in preventing the heat loss through transmission. As a matter of fact, it is common practice in the building sector to consider the level of thermal insulation of the envelope as the first indicator for describing the energy quality of the building.

Actually, the warmer climatic conditions in areas of southern Europe strongly influence the building thermal energy balance in terms of cooling needs. Therefore, the glazed percentage of the building envelope and the way the openings are shaded should also be taken into account among the main indicators of the building energy quality. Another important indicator should be the thermal capacity of the construction, since it significantly affects energy performance in the case of daily variations in the direction of the thermal flux through the envelope over long periods of time (buffer effect during summer and intermediate seasons). Moreover, the useful effect of thermal mass during the warm season can even increase, depending on the way the building is ventilated (i.e. free cooling strategies): in the warmer areas, in fact, the ventilation rates usually exceed the amount strictly needed to guarantee the indoor air quality, often through window management by the users, as mechanical ventilation systems are rarely provided.

In order to properly take into account all these aspects, the assessments of the building energy balance should be performed by means of detailed evaluations (i.e. dynamic calculation procedures). In this case, however, the calculation methods are too complex to be widely adopted in common practice, even using proper simulation tools: an accurate physics model, with detailed boundary conditions defined at least on an hourly base (building usage patterns, climate, etc.), and therefore producing reliable results, can only be achieved by users with sufficient competence and experience.

Furthermore, the set-points regulating building climatization systems, defined as the "suitable indoor temperatures" providing comfortable thermal conditions, and therefore the corresponding values assumed for the building energy balance calculation, are still commonly defined according to the thermal comfort theory formulated by Fanger in the 1970s.

This approach bases the definition of thermal comfort on pure physics, neglecting social and psychological aspects of thermal perception, while in the Southern European context, and in particular in the Mediterranean region, the building solutions and users' habits reveal certain peculiarities. Buildings are in fact widely naturally ventilated even when active cooling systems are installed, because people are traditionally used to maintain contact with the outdoors, and are usually equipped with operable window shading devices. Under these conditions, the real cooling needs strongly depend on the comfort mitigation strategies adopted by the users, and the thermal expectations are strictly related to the mean outdoor climatic conditions of the considered period (thermal experience).

In this respect, the very narrow range of allowable indoor set-point temperatures defined by a completely steady-state approach for assessing building thermal energy needs is questionable.

The main alternative approach to determining, on a variable basis, comfortable environmental conditions is the adaptive approach, which has a dynamic form depending on a transient parameter, i.e. the external temperature. This approach was developed on the assumption that the occupants are an active part of the enclosed environment: i.e. they can interact with the construction and can affect the building boundary conditions regardless of the presence of an active climatization system.

Once again, however, this climatic-related indoor temperature can only be set by means of a detailed simulation analysis.

Another important aspect that can be properly taken into account only by performing a dynamic analysis is the operative temperature in the spaces, which considers the surface radiant temperatures and therefore more accurately represents the performances of the building envelope in contributing to overall thermal comfort sensation.

The operative temperature parameter can strongly influence the real building energy need, especially during summer when the solar radiation affects the glazed surfaces, and is commonly neglected in simplified building energy assessment, which considers only the air temperature. As a matter of fact, in the case of unfavourable radiant temperatures, the air temperature set-point is usually corrected by the users in order to adjust the indoor condition to compensate the radiative component, and the consequent overuse of the active climatization systems causes an unpredicted increase in building energy consumptions.

Summarizing, simplified procedures and assumptions have been common practice for a long time in assessing the energy performances of buildings in Europe: currently, the quasi-steady-state energy balance calculation method (the simplest among those provided for by EN ISO 13790) is still the main reference for implementing procedures at national level and is also widely adopted for the energy certification of buildings. Nevertheless, a proper evaluation of building energy performance, with particular reference to the southern climatic area, should be significantly more complex.

This book discusses the related issues, besides the theoretical basis, through several application case studies carried out with reference to the Italian context, considered as representative of southern Europe. These descriptions will support energy consultants and other interested parties in assessing building energy performance beyond the mere simplified standard assumptions. Furthermore, the numerous graphs and tables documenting the analysis of a set of typical building solutions can be easily adopted to serve as design tools for both new constructions and retrofits.

Finally, I hope that this book, bringing together results from some of the most significant researches that I have promoted and coordinated in recent years on the issue, will contribute to increasing awareness of the actual consequences of architectural design choices, contrasting the trends to construct excessively light, largely glazed and improperly sealed buildings, and to encouraging the definition of more suitable energy policies for the building sector in the Southern European context.

Milan
July 2015

Simone Ferrari

Contents

Chapter 1
Building Envelope and Thermal Balance

Abstract From the thermal balance point of view, in addition to air mass transfer due to infiltration and ventilation, building energy performance strictly depends on the characteristics of the envelope, in that it constitutes the boundary between the indoor and outdoor environments. Most of the currently available building performance assessment methods evaluate the heat exchange through the envelope by the means of a steady-state analysis, leading to the diffusion of strict regulations regarding the heat transmittance of the envelope elements. Although this approach is simple to use, it does not take into account the dynamic behaviour of the construction materials, which tend to store heat and release it after a certain length of time (Van Geem 1987). This phenomenon, usually referred to as thermal inertia or thermal mass, strongly affects the heat transfer process, influencing the building thermal energy need. This chapter summarises the theoretical basis of building thermal balance and heat transfer through the envelope. Furthermore, the different implications when considering the dynamic and the steady-state assessment methods are presented with the help of a practical example.

Keywords Building thermal balance · Heat transfer through building envelope · Dynamic and steady-state heat transfer · Building heat capacity · Thermal inertia · Climatic chamber tests

1.1 Building Thermal Balance

The building (or the considered thermal zone of a building) heating and cooling need calculation is mostly based on the air volume heat balance, which takes into account all the different heat flows affecting the indoor environments (Eq. 1.1). In order to simplify the calculation, the only sensible heat balance is considered, neglecting the thermal effects due to the water contained in the air (Cengel 2008).[1]

---

[1]These effects are usually taken into account only when a complete air conditioning process is considered.

S. Ferrari and V. Zanotto, *Building Energy Performance Assessment in Southern Europe*, PoliMI SpringerBriefs,
DOI 10.1007/978-3-319-24136-4_1

$$\sum \Phi_j = M \cdot c_{p,air} \frac{d\theta_{air}}{dt} \quad (1.1)$$

where

$\Sigma\Phi_j$ is the summation of the heat flow rates involving the air volume (W)
M is the air volume mass (kg)
$c_{p,air}$ is the air specific heat capacity (J/(kg K))
θ_{air} is the air volume temperature (K)
t is time (s)

The climatization system aims at maintaining a constant indoor air temperature, and so the second half of the balance usually equals zero. The summation of the sensible heat flows involving the air volume is defined in Eq. 1.2.

$$\sum \Phi_j = \Phi_{env} + \Phi_{ven} + \Phi_{\text{int}} + \Phi_{sol} + \Phi_{sys} \quad (1.2)$$

where

Φ_{env} is the heat flow rate from the envelope (W)
Φ_{ven} is the heat flow rate due to air mass exchange (W)
Φ_{int} is the heat gain due to internal heat sources (W)
Φ_{sol} is the solar heat gain through the transparent elements (W)
Φ_{sys} is the heat flow rate due to climatization system (W)

The balance is therefore solved using the heat flow rate due to the climatization system as dependent variable, as in Eq. 1.3.

$$\Phi_{sys} = \Phi_{env} + \Phi_{ven} + \Phi_{\text{int}} + \Phi_{sol} \quad (1.3)$$

The components of Eq. 1.3 have been traditionally analysed either in a detailed dynamic way or in a simplified way, which reduces the complexity of the underlying equations imposing steady-state boundary conditions. If the first approach guarantees a high accuracy level in reproducing the energy and mass flows occurring in the building, the latter one provides ease and flexibility in indicating the building performance.

1.1.1 Heat Flow from Envelope

The heat flow from the envelope is generally defined according to Eq. 1.4.

$$\Phi_{env} = \sum_{i}^{N_{el}} h_{s,i} A_i \left(\theta_{s,i} - \theta_z\right) \quad (1.4)$$

where
h_s is the surface convective and radiative heat transfer coefficient (W/(m^2 K))
A is the surface of the envelope element (m^2)
$\theta_{s,i}$ is the internal surface temperature of the envelope element (°C)
θ_z is the indoor desired temperature (°C)
N_{el} is the number of envelope elements

This parameter is essentially determined as the result of the external variability in terms of air temperature and solar radiation on the opaque elements of the envelope, which is determined by the general heat conduction equation (see Sect. 1.2).

1.1.2 Ventilation

The heat flow due to air mass exchange is defined according to Eq. 1.5.

$$\Phi_{vent} = \sum_{i}^{Nflow} \dot{m}_j c_{p,air} \left(\theta_{air,e} - \theta_z \right) \tag{1.5}$$

where
$\dot{m}$ is the air mass flow rate (kg/s)
$c_{p,air}$ is the air specific heat capacity (J/(kg K))
$\theta_{air,e}$ is the outdoor air temperature (°C)
N_{flow} is the number of air flows

The calculation of this part of the heat balance is based on the same equation both in the steady-state and dynamic approaches, except for the time-step and therefore the application schedules.

The main parameter characterizing this heat flow is the air flow rate, also defined "discharge rate", which can be either determined in detail according to opening size and presence of wind or in simplified ways according to standard air change rate values connected to the building use and provided by international and local regulations.

1.1.3 Internal Heat Sources

The heat production due to internal sources is a summation of the heat production of people, lighting and equipment present in the thermal zone. Similarly to the previously described discharge rate, also this component of the heat balance is based on the same calculation in the steady-state as well as in the dynamic approach, except for the time step and the application schedules.

The heat produced by these elements is usually summarized in overall load values, specific per floor area and provided by international and local regulations according to building use.

1.1.4 *Solar Gain Through the Transparent Elements*

The influence of the solar radiation entering the zone through the transparent elements on the indoor environment is defined according to Eq. 1.6 (Szokolay 2008).

$$Q_{sol} = A \cdot G \cdot SGF \tag{1.6}$$

where

A is the glazed surface (m^2)

G is the global (direct and diffuse) solar radiation incident on the glazed surface (W/m^2)

SGF is the solar gain factor

The solar gain factor is a decimal fraction indicating what part of the incident radiation reaches the interior, depending on the characteristics of the glass and, if provided, of the shading elements.

Once arrived on the glazing surface, some part of the radiation is transmitted, some reflected and the remainder is absorbed within the body of the glass. The absorbed part will then heat up the glass, which will emit some of this heat to the outside, some of it to the inside, by re-radiation and convection. The SGF represents the sum of this inward re-emitted heat and the direct transmission.

This component of the heat balance is based on the same calculation both in steady-state and dynamic approaches, except for the considered time step.

1.2 Heat Transfer Through Building Elements

The heat transfer through the building elements is regulated by the law of heat conduction, the Fourier's law, which states that heat transfer is directly connected to the temperature distribution inside the same element. This law takes the form of Eq. 1.7.

$$\vec{q} = -\lambda \nabla T \tag{1.7}$$

where

$\vec{q}$ is the density of heat flow rate (W/m^2)

λ is the thermal conductivity ($W/(m\ K)$)

$-\nabla T$ is the temperature gradient (K/m)

This component of the heat balance is calculated differently according to the steady-state and to the dynamic approach, as described in the following paragraphs.

1.2.1 *Steady-State Analysis*

Within the building heat balance, when dealing with the calculation of the maximum heat exchange due to reference boundary conditions (i.e. in case of the peak load needed to size the heating system), the effects of unsteady parameters such as the solar radiation and the internal heat loads can be safely neglected. The Fourier's law can be simplified considering flow crossing in steady-state conditions, without accounting for the dynamic phenomena and therefore neglecting the thermal inertia of the building elements (which depends on the density and specific heat of the materials), and in a single direction, which represents most of the heat exchanges through envelope elements (Eq. 1.8).

$$q = -\lambda \frac{dT}{dx} \tag{1.8}$$

where

q is the heat flow rate density through the layer (W/m^2)
λ is the conductivity of the layer material (W/(m K))
dT is the temperature difference of the layer surfaces (K)
dx is the penetration depth of the heat flow (m)

With constant heat flow rate and penetration depth, which corresponds to the thickness of a homogeneous plane layer, thermal resistance R can be defined according to Eq. 1.9.

$$R = \frac{\Delta x}{\lambda} = \frac{s}{\lambda} \tag{1.9}$$

where

R is the thermal resistance of the homogeneous plane wall ((m^2 K)/W)
s is the plane wall thickness (m)

From this quantity it was possible to develop a model analogue to the one of electrical connections. This allows therefore calculating the overall resistance of a multilayer wall as the sum of the resistances of the n homogeneous layers that are part of it, taken in series, as in Eq. 1.10.

$$R_{wall} = \sum_{i=1}^{n} R_{layer,i} \tag{1.10}$$

Simplifying, then, in case of steady-state heat conduction an overall parameter characterizing the heat transfer through a multilayer wall can be used: the thermal conductance (Eq. 1.11).

$$C = \frac{1}{R_{wall}} \tag{1.11}$$

Finally, in order to take into account the surface heat transfer due to convection and radiation towards the surrounding air, the surface resistance value was also defined (Eq. 1.12). It depends partly on the surface characteristics and partly on the environmental conditions of the surrounding air.

$$R_s = \frac{1}{h} \tag{1.12}$$

where
h is the surface heat transfer coefficient (W/(m^2 K))

Combining the conduction thermal characteristics with the superficial transfer coefficients it was possible to define an overall quantity describing the steady-state heat transfer characteristics of the complete element, which is the thermal transmittance, or U-value (W/(m^2 K)).

$$U = \left(R_{s,i} + R_{wall} + R_{s,e}\right)^{-1} = \left(\frac{1}{h_e} + \sum_{i=1}^{n} \frac{s_i}{\lambda_i} + \frac{1}{h_i}\right)^{-1} \tag{1.13}$$

The calculation of thermal resistance and U-value for the envelope elements (Eq. 1.13) is described in norm EN ISO 6946 (2007), which also provides standard reference values for the surface resistances according to the element type.

Therefore, in case of heat exchange through the building envelope due to simple transmission in steady-state conditions Eq. 1.14 may be used when aiming at determining the winter peak load.

$$\Phi_{env,tr} = \sum_{i}^{N_{el}} U_i A_i \left(T_{air,e} - T_z\right) \tag{1.14}$$

where
U is the thermal transmittance of the element (W/(m^2 K))
A is the surface of the envelope element (m^2)

$T_{air,e}$ is the external air temperature (°C)
T_z is the internal desired temperature (°C)

This simplified approach was also extended to the calculation of the heat balance performed for longer time intervals (monthly or seasonal) in order to determine the building climatization energy need. In Eq. 1.14, however, only the heat flow caused by the temperature variation is taken into account, while the heat contribution due to the solar radiation arriving on the opaque envelope element is neglected. This approach is suitable for the winter peak load assessment, but in case of the energy need the solar component is very important and is usually added to the heat balance equation through Eq. 1.15.

$$\Phi_{env,sol,op} = \sum_{j}^{N_{el,op}} R_{se,j} U_j A_j \left(F_{sh,j} \alpha_j I_{sol} - h_{r,j} \Delta T_{er}\right) \quad (1.15)$$

where

$R_{s,e}$ is the thermal resistance of the external surface ((m^2 K)/W)
U is the thermal transmittance of the element (W/(m^2 K))
A is the surface of the element (m^2)
F_{sh} is the shading factor due to the surroundings
A is the surface absorption coefficient
I_{sol} is the solar irradiance (W/m^2)
h_r is the radiative heat transfer coefficient of the external surface (W/(m^2 K))
ΔT_{er} is the average difference between air temperature and apparent sky temperature (K)
$N_{el,op}$ is the number of opaque envelope elements

Often Eqs. 1.14 and 1.15 are combined, grouping the *UA* parameter and adjusting the temperature difference parameter by substituting the outdoor air temperature with the sol-air one, which is defined according to Eq. 1.16 (ASHRAE 2001).

$$T_{sa,e} = T_{air,e} + \frac{\alpha I}{h_e} - \frac{\varepsilon \Delta R}{h_e} \quad (1.16)$$

where

$T_{sa,e}$ is the sol-air temperature (°C)
$T_{air,e}$ is the outdoor air temperature (°C)
α is the absorption coefficient of the external surface of the envelope element
I is the global solar radiation on the envelope element (W/m^2)
ε is the emissivity of the external surface of the envelope element

ΔR is the difference between the infrared radiation from the surroundings of the building (included the sky) and the one emitted by a black body at the outdoor air temperature (W/m^2)

h_e is the external surface heat transfer coefficient (W/(m^2 K))

1.2.2 *Transient Analysis*

If the Fourier's equation is combined with the energy conservation principle for a minimal volume, it becomes the general heat conduction equation (Eq. 1.17).

$$\rho c \frac{\partial T}{\partial t} = \lambda \left[\frac{\partial^2 T}{\partial x^2} + \frac{\partial^2 T}{\partial y^2} + \frac{\partial^2 T}{\partial z^2} \right] \tag{1.17}$$

where

ρ is the material density (kg/m^3)

c is the material specific heat (J/(kg K))

∂T is the temperature variation (K)

∂t is the time variation (s)

λ is the material conductivity (W/(m K))

x,y,z represent the space coordinates

As done for the steady-state analysis, also in the transient case the heat flow through building envelope elements can be approximated to a one-dimensional formulation and so the general heat conduction equation can be expressed as in Eq. 1.18.

$$\rho c \frac{\partial T}{\partial t} = \lambda \frac{\partial^2 T}{\partial x^2} \tag{1.18}$$

Equation 1.18 is a partial differential equation (PDE) and therefore it is not easily solvable outside specific boundary conditions (i.e. periodic analysis). The common way to face this equation is to adopt numerical methods that approximately solve it developing a function (or a discretization of this function) which simultaneously satisfies a certain relation between its derivatives calculated in specific space or time regions and some given boundary conditions at the extremes of the domain. Even if these methods simplify the calculation procedure, they require strong engineering knowledge to use the complicated application software.

1.2.2.1 Analysis Through Discretization

The main numerical methods that exactly solve PDEs are the finite difference, finite volumes and finite element analysis, which are based on an approximation of the reference domain in order to simplify the solution of the heat conduction equation.

The finite difference method is based on the reduction of the derivatives to a system of equations expressing the difference between the values of the function in discrete points.

The finite volume method is equally based on the values calculated in discrete points, but these points are found within a geometric mesh, whose vertices are placed inside infinitesimal volumes, and the heat flows are calculated at their boundaries (faces and edges). This technique is very common to solve fluid-dynamic problems.

Finally, the finite element method was developed in 1950s in order to face the complexity of some elasticity and structural problems, and is based on the simplification of the domain by its discretization in subdomains, which are geometric primitives (mostly triangles in case of two-dimensional domains and tetrahedra in case of three-dimensional domains) and are called "elements". In this way it is possible to model objects with a very wide range of geometries, by the means of a mesh which appropriately discretizes the domain, a detailed description of all the relevant physical characteristics, and the description of the boundary conditions that activate the phenomenon.

As previously said, in case of the heat transfer through an envelope element a single direction flow is considered accurate enough, and so a wall can be easily modeled as its cross section, with all the different layers described by their thermal characteristics. The boundary conditions activating the heat exchange can be described directly as density of heat flow rate on the wall surfaces, or as a combination of air temperature coupled with a surface (radiative and convective) heat transfer coefficient, or even as a forced temperature for the wall surface.

1.2.2.2 Conduction Transfer Functions

The discretizing methods are complex to solve, and generally need the recourse to calculation means that have become available only recently. Therefore it was developed a simplified numerical method based on the "transfer functions" method, which represents systems by the means of an impulse-result scheme. In the field of heat transfer, this method took the name of conduction transfer functions or CTFs, and was developed in the 1960s by G.P. Mitalas and D.G. Stephenson from National Research Council of Canada (NRCC) and accepted by the American Society of Heating, Refrigerating, and Air-Conditioning Engineers (ASHRAE), which included it in its Handbook of Fundamentals. For this reason this technique is also known as "ASHRAE method".

The transient nature of both inputs and outputs of the system is often faced using mathematical transforms. Among the most common ones regarding heat transfer there are the Laplace transform for the analysis of continuous domains and the Z-transform, which was developed within the signal processing, for the analysis of data within a discrete domain.

Conduction transfer functions portray the whole dynamic behaviour of a building element and are a feature of the system depending solely on the thermo-physical characteristics of its single homogeneous layers. According to the boundary conditions at the element surface (i.e. the temperature oscillation) it is possible to determine the conduction heat flows through the boundary layers, characterized by unitary surface heat transfer coefficients (Eq. 1.19).

$$\begin{cases} \varphi_e(z) = \dfrac{D(z)}{B(z)} \cdot T_e(z) - \dfrac{1}{B(z)} \cdot T_i(z) \\ \varphi_i(z) = \dfrac{1}{B(z)} \cdot T_e(z) - \dfrac{A(z)}{B(z)} \cdot T_i(z) \end{cases} \tag{1.19}$$

where

$\varphi(z)$	is the Z-transform of heat flow rate density
$T(z)$	is the Z-transform of temperature
$D(z)/B(z)$	is a conduction transfer function through the element
$A(z)/B(z)$	is a conduction transfer function through the component
$1/B(z)$	is a conduction transfer function through the component

Therefore, CTFs depend on:

- the thermo-physical characteristics of the layers materials (homogeneous or alike) in the element;
- the position of the layers within the element;
- the boundary conditions described by the surface heat transfer coefficients;
- the analysis time interval (usually 1 h).

Conduction transfer functions are generally determined according to how specific construction elements react to specific thermal stimuli (i.e. step, linear ramp, parabolic ramp).

The CTF method is the most common among the building dynamic simulation software since it has a relatively high computation speed (functions can be calculated once for each construction element at the beginning of the simulation) and it is generally accurate. Among the most famous tools for evaluating the transient energy balance of the buildings there are the DOE and TRNSYS (Klein et al. 2007) codes, based on the Z-transform according to the ASHRAE method, the BLAST one, based directly on the Laplace transform, and Energy Plus (Crawley et al. 2001), which implements a procedure called “state space method” which is based on several equations in matrix form.

1.2.2.3 Periodic Analysis

In case of specific problems, which are characterized by simple geometry, simple boundary conditions and material properties that are homogeneous and independent from temperature, PDEs also allow analytical solutions. In case of heat transfer, as alternative to both the complex numerical methods and the excessively simplified steady-state method, periodic analysis was therefore introduced, assuming that the envelope elements undertake a sinusoidal temperature variation.

A lot of physical phenomena can be approximated to sinusoidal trends, in particular when considering the thermal stimuli of buildings (i.e. the daily or yearly trends of climatic parameters): therefore some properties of periodic functions can reasonably be extended to portray the heat conduction through envelope elements in a simplified way. International standard EN ISO 13786 (2007), in fact, specifies the characteristics that describe the thermal behaviour of the envelope elements under periodic boundary conditions and suggests a procedure to calculate them. This norm assumes a sinusoidal temperature variation on one side of the element (usually the outside environment), while on the other side (usually the inside space) temperature is held constant. The effect of this stimulus is reflected on the heat flows through the element surfaces, which are one dimensional and have sinusoidal trends as well.

Within the standard, the following heat transfer matrix for a layer Z is defined (Eq. 1.20)

$$Z = \begin{pmatrix} \hat{\theta}_2 \\ \hat{q}_2 \end{pmatrix} = \begin{pmatrix} Z_{11} & Z_{12} \\ Z_{21} & Z_{22} \end{pmatrix} \cdot \begin{pmatrix} \hat{\theta}_1 \\ \hat{q}_1 \end{pmatrix} \tag{1.20}$$

where

$\hat{q}$ is the complex amplitude of the heat flow rate density through the surface of the element (W/m^2)

$\hat{\theta}$ is the complex amplitude of the air temperature (K)

Z_{ij} is an element of the transfer matrix, defined by Eq. 1.21

$$Z_{ij} = f(T, \lambda, \delta, \xi) \tag{1.21}$$

where

T is the period of the variation (s)

λ is the conductivity of the layer material (W/(m K))

δ is the periodic penetration depth of the heat wave in the layer (m)

ξ is the ratio of the thickness of the layer to the penetration depth

The transfer matrix for a multi-layer element results from the multiplication of the matrices characterizing the single layers.

This method can be adopted to solve any problem connecting the boundary conditions on one side of the element and the heat flow rate on the other side, consistently with the method assumptions.

Moreover, once the transfer matrix is defined it is possible to calculate some dynamic properties of the considered construction, representing the response of the element to a sinusoidal temperature variation on one of its sides.

Thermal admittance (Eq. 1.22), for instance, represents the complex amplitude of the heat flow rate density through the surface of the component divided by the complex amplitude of the temperature variation in the zone on the same side of the element (m).

$$Y_{mm} = \frac{\hat{q}_m}{\hat{\theta}_m} \quad (1.22)$$

where

Y_{mm} is the thermal admittance (W/(m^2 K))

$\hat{q}_m$ is the complex amplitude of the heat flow rate density through the element surface adjacent to zone m (W/m^2)

$\hat{q}_m$ is the complex amplitude of the air temperature within zone m (W/m^2)

Periodic thermal transmittance (Eq. 1.23), differently, represents the complex amplitude of heat flow rate density through the element surface on the side with steady conditions, divided by the complex amplitude of the temperature variation on the other side.

$$Y_{mn} = \frac{\hat{q}_m}{\hat{\theta}_n} \quad (1.23)$$

where

Y_{mn} is the periodic thermal transmittance (W/(m^2 K))

$\hat{q}_m$ is the complex amplitude of the heat flow rate density through the element surface adjacent to zone m (W/m^2)

$\hat{\theta}_n$ is the complex amplitude of air temperature within zone n (W/m^2)

Thermal admittance and periodic thermal transmittance are therefore complex numbers as well, and the value usually taken into account for periodic analysis calculation is the modulus. In case of low thermal inertia both moduli tend to the element U-value.

From these fundamental characteristics it is also possible to determine the decrement factor (Eq. 1.24), which represents the dampening of the heat flow oscillation when crossing the element.

$$f = \frac{|Y_{mn}|}{U} \tag{1.24}$$

where
Y_{mn} is the periodic thermal transmittance (W/(m^2 K))
U is the U-value (W/(m^2 K))

As can the *time shift* (Eq. 1.25), which represents the delay of the heat flow in crossing the element:

$$\Delta t_f = \frac{T}{2\pi} \arg(Y_{mn}) \tag{1.25}$$

where
T is the period of the variation (s)

Additionally, these phenomena can also be interpreted in terms of the ratio between the temperature oscillation on the outside and the temperature response on the inside surface facing a room at constant temperature (Ulgen 2002; Kaşka and Yumrutaş 2009), giving the Decrement Factor (*DF*) and the Time Lag (*TL*, in h) definition described through Eqs. 1.26 and 1.27 respectively.

$$DF = \frac{T_{s,i,\max} - T_{s,i,\min}}{T_{sa,e,\max} - T_{sa,e,\min}} \tag{1.26}$$

and

$$TL = t_{T_{s,i,\max}} - t_{T_{sa,e,\max}} \tag{1.27}$$

where
$T_{s,i}$ inside surface temperature, in °C
$T_{sa,e}$ external sol-air temperature, in °C
min local minimum
max local maximum
t time instant when the local maxima take place, in h

Among the properties described in the standard, the ones which are easier to understand and use are the time shift and the decrement factor. Moreover, the thermal admittance value is considered in standard EN ISO 13792 (2005), which suggests a simplified procedure to calculate the summer indoor temperature of a room without air-conditioning system ("admittance procedure").

Another parameter introduced by the EN ISO 13786 (2007) standard is the areal heat capacity, which represents the effective heat capacity, relative to the layers directly interacting to the temperature variation on one of the side of the element

(Eq. 1.28). Therefore, for each of the building envelope elements there can be calculated both an internal areal heat capacity and an external one.

$$\kappa_m = \frac{T}{2\pi} |Y_{mm} - Y_{mn}| \tag{1.28}$$

where

κ_m is the aeral heat capacity of the m-facing side of the element (J/(m^2 K))
T is the period of the variation (s)
Y_{mm} is the thermal admittance (W/(m^2 K))
Y_{mn} is the periodic thermal transmittance (W/(m^2 K))

Standard EN ISO 13790 (2008) refer to this last characteristic in the calculation of the utilization factors in order to take into account a dynamic parameter in the simplified steady-state building energy balance (see Chap. 3). In Standard EN ISO 13786 (2007) it is also possible to find some simplified methods to calculate the aeral heat capacity without using the transfer matrix.

1.2.3 Steady-State Versus Transient Prediction

Within a larger study carried out in Politecnico di Milano (Ferrari and Zanotto 2013), it was possible to compare the actual behaviour of four selected walls, measured by the means of climatic chamber tests (ASTM C1363 2005; Brown and Stephenson 1993a, b; Burch et al. 1990; ISO 8890 1994), and the performance of the same walls assessed by the means of a finite element analysis tool.

Since the main dynamic phenomenon that can be seen in walls is its thermal inertia, the selected constructions are characterized by the same U-value but by different aeric mass and layout (i.e. insulation position):

- a heavyweight masonry wall (“Heavy”), made by one layer of high density hollow bricks;
- a medium/heavyweight wall (“Mid-Ins”), made by outdoor face bricks and a hollow brick layer with insulation within;
- a medium/heavyweight wall (“Ext-Ins”) with hollow bricks and exterior insulation;
- a lightweight wall (“Light”) composed by an insulation sandwich panel.

The characteristics for each of these wall types are summarised in Table 1.1.

The walls were tested by the means of daily cycles of sol-air temperature (ASHRAE 2001), in order to represent a likely thermal impulse on a vertical wall facing South located in the city of Rome (considered as average representative of the Italian context). Two days were chosen from the Test Reference Year (TRY) climatic data of Rome: a very cold winter day (January 26th) and a very warm summer day (July 17th) without significant perturbation by occasional meteorological events.

Table 1.1 Chosen wall types and related main thermal characteristics

	Heavy	Mid-ins	Ext-ins	Light
Thickness (m)	0.480	0.365	0.400	0.125
U-value (W/(m^2 K))	0.298	0.299	0.306	0.312
Aeric mass (kg/m^2)	431	343	301	35
Heat capacity (kJ/(m^2 K))	368	291	259	28

For each sample, the test lasted at least 3 days, with a repetition of the daily cycle temperatures inside the climatic chamber, in order to stabilize the heat flows.

Regarding the finite element analysis, the wall samples have been modelled as 2 m long specimens (two-dimensional domains) and the analysis points have been placed in the middle, in order to reflect the sensors location for the climatic chamber tests.

The boundary conditions imposed to the domain are the surface temperatures registered during the tests, in order to neglect the variability connected to the heat transfer coefficient, which is difficult to estimate and strongly depends on the instantaneous environmental conditions. The simulation has been performed with the same boundary settings for a 30 days' time in order to stabilize the model, in particular regarding the heavier samples.

The temperatures predicted by the finite element method for the internal layers are very similar to the ones measured during the actual tests, both regarding the general trend and the absolute numeric values: the maximum differences are detected in the medium-weight solution with external insulation during the winter day, regarding the sensor placed between the hollow bricks and the insulation, but it is always less than 3 K, as can be seen in Fig. 1.1 ($T_{s,2}$).

These small differences can be easily explained considering:

- the model approximation, which represents the hollow bricks as a single homogeneous layer with average equivalent density (consistently with the common practice in building simulation);
- since the tested samples were hand-crafted, the low precision of the actual construction praxis, in particular regarding plaster and mortar thicknesses;
- the moisture effects on the heat transfer, which were neglected in the finite element model aiming at calculation simplicity;
- three-dimensional heat transfer effects inside the real samples.

The finite element method shows, anyway, a very high accuracy in predicting the behaviour of the walls, both regarding the absolute values and the general trends.

As further analysis, the same model was used with different boundary conditions to simulate an indoor environment characterised by controlled temperature. The

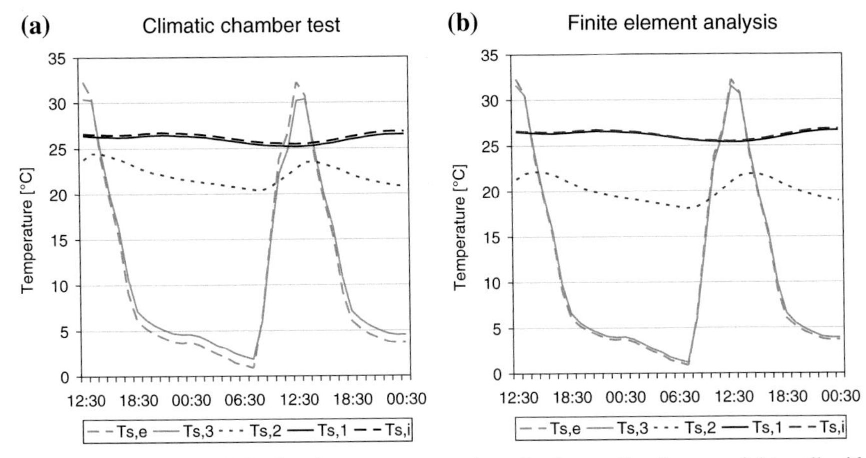

Fig. 1.1 Comparison of the hourly temperature values for the medium/heavyweight wall with external insulation during the winter day as detected by the thermocouples (**a**) and the ones calculated by the finite element simulations (**b**)

boundary conditions are based on the hourly sol-air temperature and the standard external heat transfer coefficient (25 W/(m^2 K), according to EN ISO 6946 2007) on the outside surface, and due to constant temperature (20 °C for the winter season and 26 °C for the summer season, according to EN 15251 2007) and the standard internal heat transfer coefficient (7.7 W/(m^2 K), according to EN ISO 6946 2007) on the inside surface.

In this case the simulation is extended to the whole heating season, which for Rome lasts from the November 1st to April 15th (D.P.R. 412 1993), and the whole cooling season, lasting from July 1st to the September 15th (selected as the whole period when the average daily outside sol-air temperature exceeds the comfort limit of 26 °C). Also for this analysis a stabilization time of 30 days has been respected.

Figure 1.2 shows the hourly trends regarding the inside surface temperature for the different wall types during the reference winter (January 26th) and summer (July 17th) days.

As the inside air temperature is a constant value, a visible oscillation of the surface temperature is only detectable in case of the lightweight solution. This variability exists even if the air with constant temperature is considered directly in contact with the surface, attenuating the actual oscillation effect which would occur in a real room, where the air temperature is not evenly distributed.

Acknowledging this behaviour, the attenuation and time delay phenomena, as described by Eqs. 1.25 and 1.26 for the periodic analysis, are calculated according to the finite element simulation results.

The outcoming values are shown in Fig. 1.3 and show the same general trend as the ones resulting from the experimental tests, but the differences between the lightweight construction and the others increase: regarding the time lag, the discrepancies among the various heavyweight samples can be seen, too. The summer

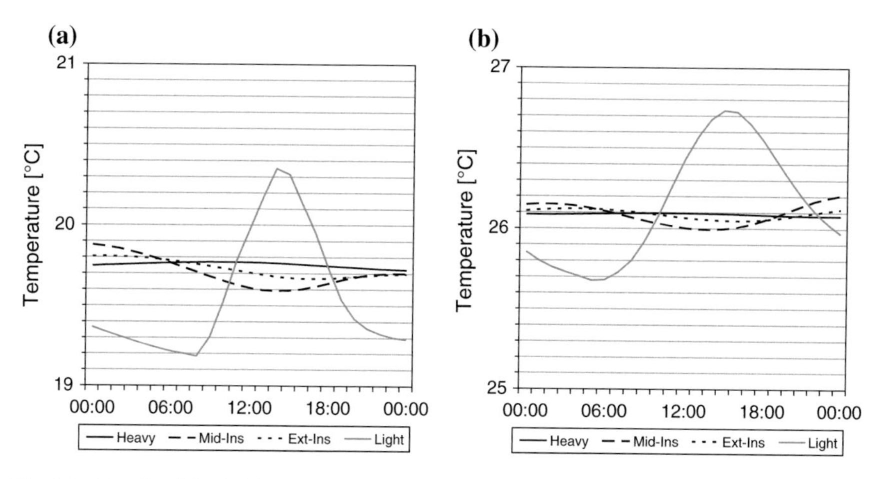

Fig. 1.2 Trends of the inside surface temperature as calculated for the different wall types in the reference winter day (**a**) and in the reference summer day (**b**)

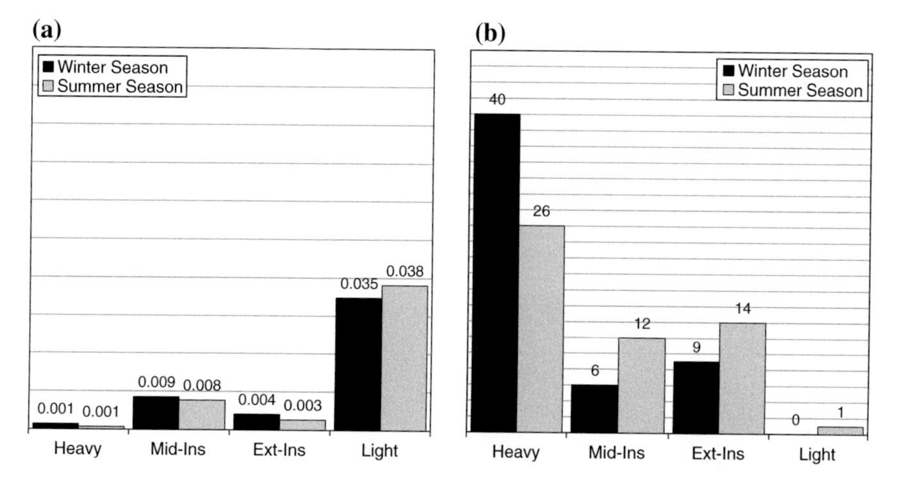

Fig. 1.3 Decrement factor (*DF*) (**a**) and time lag (*TL*) (**b**) results for the different wall samples

season thermal delay values, in particular, are very similar to the ones resulting from the nominal periodic parameters. This can be explained considering that the summer solicitation is very close to the sinusoidal trend assumed in EN ISO 13786 (2007).

The finite element analysis allows to take into account not only the temperature results, but also the density of heat flow rate values crossing the analysis points, in W/m^2. In this work, the seasonal peak load has been estimated as the maximum values registered during the winter and summer seasons, as shown in Table 1.2.

Differently, with the conventional simplified approach as reference, the peak load should be the same for all the walls, since it is calculated as the product of the nominal U-value and the seasonal maximum difference between the indoor air and the external sol-air temperatures.

The differences between the two evaluations are pointed out in Fig. 1.4 and show how the lightweight sample, with low thermal inertia, behaves closely to the steady-state estimation (with percentage near 100 %), while the other solutions strongly attenuate the effects of the sol-air temperature. It is useful to remind that in case of the heavyweight walls the peak load does not occur directly after the maximum solicitation, but only after a time lag corresponding to the thermal delay (the one shown in Fig. 1.3).

As overall results, the finite element simulation lead to various remarks in the field of energy performance implications, demonstrating the importance of correctly taking into account the dynamic behaviour of the building envelope.

Table 1.2 Peak load values for the density of heat flow rate, in W/m^2, calculated by the means of the finite element analysis during the winter and summer season according to the different constructions

	Heavy	Mid-ins	Ext-ins	Light
Winter season	3.37	4.24	3.84	6.29
Summer season	1.69	2.64	1.99	6.86

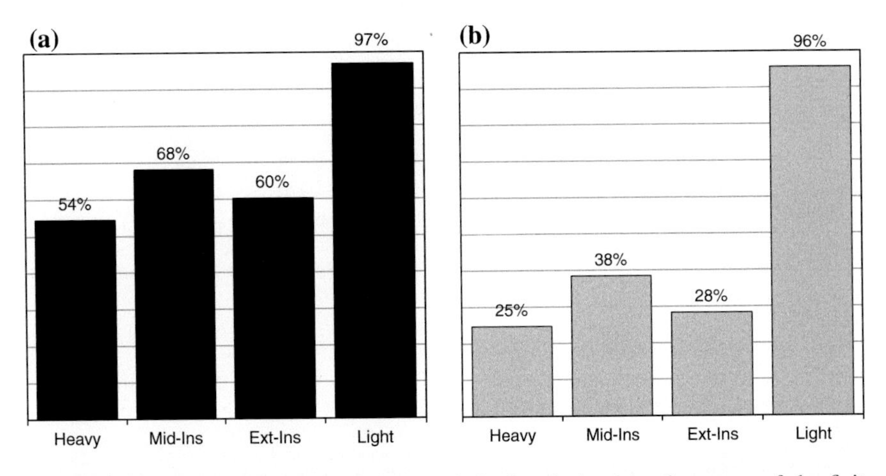

Fig. 1.4 Ratio (in percentage) between the peak loads calculated by the means of the finite element analysis and the ones estimated by the means of the steady-state approach in winter (**a**) and summer (**b**)

First of all, the attenuation of the heat wave oscillation represented by the temperature decrement factor can be interpreted as:

- lower peak load which would bring to a smaller size for the climatization system;
- smaller variability of the internal surface temperature, which means a more even use of the heating and cooling system (on/off operation) and more constant environmental conditions, influencing thermal comfort (radiant temperature).

The thermal delay, on the other hand, can bring to clear advantages on the indoor environmental conditions in the summer season, since the inside surface temperature (and therefore also the related heat exchange) has its local maxima during the night time, when most of the internal heat loads do not affect the zone volume and when the outdoor environmental conditions allow the use of attenuation and passive cooling strategies (i.e. night ventilation).

References

ASHRAE, *Handbook of Fundamentals,* (American Society of Heating, Refrigerating and Air-conditioning Engineers, Atlanta, 2001)

ASTM C1363, *Standard Test Method for Thermal Performance of Building Materials and Envelope Assemblies by Means of a Test Box Apparatus,* (American Society for Testing and Materials, West Consohoken, PA, 2005)

W.C. Brown, D.G. Stephenson, Guarded hot box measurements of the dynamic heat transmission characteristics of seven wall specimens—Part I. ASHRAE Trans. **99**(1), 632–642 (1993a)

W.C. Brown, D.G. Stephenson, Guarded hot box measurements of the dynamic heat transmission characteristics of seven wall specimens—Part II. ASHRAE Trans. **99**(1), 643–660 (1993b)

D.M. Burch, B.A. Licitra, R.R. Zarr, A comparison of two test methods for determining transfer function coefficients for a wall using a calibrated hot box. J. Heat Transf. **112**, 35–42 (1990)

Y.A. Cengel, *Introduction to Thermodynamics and Heat Transfer* (McGraw Hill, New York, 2008)

D.B. Crawley, L.K. Lawrie, F.C. Winkelmann et al., Energyplus: creating a new generation building energy simulation program. Energy Build. **33**, 319–331 (2001)

D.P.R. 412, *Regolamento recante norme per la progettazione, l'installazione e la manutenzione degli impianti termici degli edifici, ai fini del contenimento dei consumi di energia, in attuazione del'art. 4, comma 4 della legge 9 gennaio 1991, n.10. (aggiornata dal D.P. R.551/99)* (Rome, 1993)

EN 15251, *Indoor Environmental Input Parameters for Design and Assessment of Energy Performance of Buildings Addressing Indoor Air Quality, Thermal Environment, Lighting and Acoustics* (European Committee of Standardization, Brussels, 2007)

EN ISO 6946, *Building Components and Building Elements—Thermal Resistance and Thermal Transmittance—Calculation Method* (European Committee of Standardization, Brussels, 2007)

EN ISO 13786, *Thermal Performance of Building Components—Dynamic Thermal Characteristics—Calculation Methods* (European Committee of Standardization, Brussels, 2007)

EN ISO 13790, *Energy Performance of Buildings—Calculation of Energy Use for Space Heating and Cooling* (European Committee of Standardization, Brussels, 2008)

EN ISO 13792, *Thermal Performance of Buildings—Calculation of Internal Temperatures of a Room in Summer without Mechanical Cooling—Simplified Methods* (European Committee of Standardization, Brussels, 2005)

S. Ferrari, V. Zanotto, The thermal performance of walls under actual service conditions: evaluating the results of climatic chamber tests. Constr. Build. Mater. **43**, 309–316 (2013)

ISO 8990, *Thermal Insulation—Determination of Steady-State Thermal Transmission Properties —Calibrated and Guarded Hot Box* (International Organization for Standardization, Geneva, 1994)

Ö. Kaşka, R. Yumrutaş, Experimental investigation for total equivalent temperature difference (TETD) values of building walls and roofs. Energy Convers. Manage. **50**(11), 2818–2825 (2009)

S.A. Klein, W.A. Beckman, J.W. Mitchell, et al., *TRNSYS—A Transient System Simulation Program User Manual* (The solar energy Laboratory—University of Wisconsin, Madison, 2007)

S.V. Szokolay, *Introduction to architectural science: the basis of sustainable design* (Architectural Press-Elsevier Ltd., Amsterdam, 2008)

K. Ulgen, Experimental and theoretical investigation of effects of wall's thermophysical properties on time lag and decrement factor. Energy Build. **34**, 273–278 (2002)

M.G. Van Geem, Thermal mass: what the R-values neglect. Constr. Specif. **1987**, 71–77 (1987)

Chapter 2
Approximating Dynamic Thermal Behaviour of the Building Envelope

Abstract Assessing the building energy performance through the simplified assumptions of the steady-state conditions can prove restrictive. The estimation of heat transfer through the envelope is based on the U-value of the building construction, neglecting the thermal mass effect. Several approaches have, therefore, been developed to approximate the dynamic behaviour of buildings when performing a standard steady-state analysis. According to Barnaby (1982), these methods can be generally divided into those analysing an isolated building element and those considering the context of whole building. The first category introduces some correction values that affect the main parameters of the steady-state heat transmission calculation (i.e. the U-value of the envelope elements and the temperature difference between the indoor and outdoor environments). This chapter investigates a set of correction values that try to represent the dynamic thermal envelope performance in a simplified manner, applying some of them to a set of case-study wall types.

Keywords Heat transfer through building envelope · Thermal inertia · Dynamic heat transfer · Correction values approximating dynamic behaviour

2.1 Heat Transmittance Correction Values

Some of the simplified methods approximating the dynamic behaviour of the building envelope starting from the steady-state approach apply a correction to the U-value of the considered element and are usually referred to the heating season.

2.1.1 Mass Factor

The M factor (or mass factor) is an adjustment value for the heat transmittance of envelope elements developed by Yu (Hankins and Anderson 1976; Yu 1978), which applies to the heating season calculation.

S. Ferrari and V. Zanotto, *Building Energy Performance Assessment in Southern Europe*, PoliMI SpringerBriefs,
DOI 10.1007/978-3-319-24136-4_2

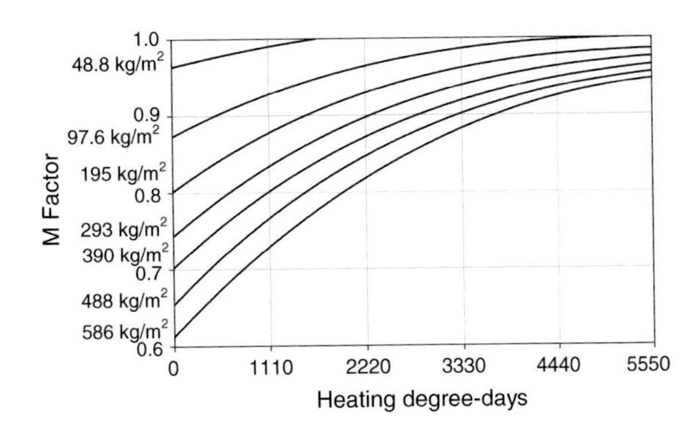

Fig. 2.1 Correlation graph for the M factor, the heating degree-days and the element frontal mass [kg/m^2] in SI units

The correction factor was defined as the ratio of the heat flow for a given wall calculated by the means of a dynamic simulation (assumed as average in January at 8:00 am) and the one calculated by the means of the steady-state method. Several typical envelope constructions were tested and used to develop reference values, which correlate the mass factor both to the element frontal mass and to the heating degree-days of the building location. These values were later translated into a graph in order to enlarge the method's applicability (Fig. 2.1).

This approach did not find large application regarding the heating energy need assessment, since several later studies found it unreliable in this matter because of it extending a value originally calculated for a specific hour to the whole season (Godfrey et al. 1979). In fact, in some cases M factor has been suggested in order to predict the savings due to envelope mass regarding the heating peak load calculation.

2.1.2 *Effective U-Value*

The effective U-value method was developed by Van der Meer (1978). This parameter is defined as the ratio of the seasonal average density of heat flow rate and the seasonal average temperature difference between the indoor and the outdoor environments. By the means of an ad hoc developed software, Van der Meer performed series of dynamic energy simulations for a whole set of construction technologies in relation to the New Mexico climatic conditions, and derived reference values for the most common wall constructions. The aim of this parameter is not to determine a comparison between different constructions, but to make a correction of the traditional U-value and to determine how the element behaviour changes as its orientation or its surface colour are modified. Later on values of this parameter have been calculated for a wider range of construction and they have been implemented in the New Mexico Building Code. As for other simplified methods, the effective U-value is not universally applicable since it is based on a list of reference values that were calculated, by the means of detailed dynamic simulations, for specific kinds of envelope technologies and specific climatic conditions.

2.2 Temperature Difference Correction Values

Some other methods apply a correction value to the temperature difference needed to calculate the heat transfer through the building envelope. They are usually referred to the cooling season.

2.2.1 *Total Equivalent Temperature Differential (TETD)*

Since 1940s, the American Society of Heating, Refrigerating and Air-Conditioning Engineers (ASHRAE) started developing a method to calculate the thermal gains for the peak cooling load estimation through the opaque envelope based on an alternative value of the temperature difference, which could take into account the combined effects of solar radiation and envelope thermal inertia. Starting from experimental data and calculation using the periodic analysis (see Chap. 1), some default data regarding specific envelope technologies were derived as the ratio of the calculated heat flow and the nominal U-value and were collected into reference tables. In ASHRAE (1972) the total equivalent temperature differential (TETD) was introduced to determine the cooling peak load. The method can be applied both choosing pre-calculated values from the reference tables, which were calculated for specific constructions and for mild climates, and calculating directly the specific value by the means of a general equation that depends mostly on the decrement factor and on the time lag of the analysed building envelope (Eq. 2.1).

$$TETD_t = T_{as,e,t} - T_{a,i,t} + f\left(T_{as,e,t-\delta} - T_{as,e,t}\right) \tag{2.1}$$

where

$T_{as,e}$ outdoor sun-air temperature [K]

$T_{a,i}$ indoor air temperature [K]

f decrement factor

δ time lag [h]

Because of this last option, this method can be considered simple and at the same time accurate in calculating the cooling load. In fact it can still be found in current researches, even if with more sophisticate calculations of the time lag and decrement factor values (Yumrutaş et al. 2007; Kaşka and Yumrutaş 2009).

2.2.2 *Cooling Load Temperature Difference (CLTD)*

Another method, which is still suggested by ASHRAE (2001) as a simplified assessment method for residential buildings cooling load calculation, is the Cooling Load Temperature Difference (CLTD).

Also in this case, standard reference values for typical envelope constructions (the basic ones considered by ASHRAE and classified in different categories), with specific U-values and frontal mass, were derived by the ratio of the dynamic heat flow, calculated by the means of the conduction transfer function (CTF) method, and the nominal U-value. The boundary conditions for the dynamic calculation were:

- 40° north latitude (which influences the solar radiation effect);
- July as the standard month for the cooling load evaluation;
- 29.4 °C of outdoor design temperature;
- 25.5 °C of indoor design temperature;
- a dark coloured outside surface.

As for most of these simplified methods, the accuracy of the provided values strongly depends on the similarities between the actual building and the simulation conducted by ASHRAE: in order to address this problem, in ASHRAE (1989) Eq. 2.2 was proposed to adjust the table data for boundary parameters different from the previously mentioned ones and for the presence of additional insulation.

$$CLTD_{corr} = (CLTD + LM)K + \left(25.5 - T_{set,i}\right) + \left(T_{a,e,prj} - 29.4\right) \tag{2.2}$$

where

$CLTD$ is the standard CLTD for typical boundary conditions [K]
LM is the adjustment factor for latitude, reference month and orientation [°C]
K is the adjustment factor for surface finishing [%]
$T_{set,i}$ is the internal set-point temperature [°C]
$T_{a,e,prj}$ is the outside air design temperature [°C]

Later researches still proposed the adoption of this method, trying to develop data for a wider range of boundary conditions. Bansal et al. (2008), in particular, calculated reference CLTD values for typical constructions and climate of India, solving the Fourier's equation by the means of the finite difference method.

2.2.3 *Overall Thermal Transfer Value (OTTV)*

As a representation of the energy consumption due to the building envelope, always in summer, ASHRAE developed the overall thermal transfer value (OTTV), which tries to combine the effect of the envelope elements exposed to different orientations and of both the opaque and transparent parts of these elements.

$$OTTV = \frac{(A_{op} \cdot U \cdot \alpha \cdot TD_{EQ}) + (A_{win} \cdot SC \cdot ESM \cdot SF)}{A} \tag{2.3}$$

where

A	envelope surface [m^2]
A_{op}	surface of the opaque part of the envelope [m^2]
U	thermal transmittance of the opaque part of the envelope [W/(m^2 K)]
α	absorbtance of the opaque part of the envelope
TD_{EQ}	equivalent temperature difference for the opaque part of the envelope [K]
A_{win}	surface of the glazed part of the envelope [m^2]
SC	shading coefficient of the surface
ESM	external shading multiplier (depending on the orientation)
SF	solar factor of the glazed surface [W/m^2]

Considering a single opaque element, the characteristics taken into account by this index are the U-value, the thermal absorptance and an adjusted value of temperature difference. The reference values of equivalent temperature difference were listed according to the surface orientation and to the element weight, after being calculated by the means of dynamic simulations. Even if since 1989 this index is not part of the ASHRAE Standard 90.1 anymore, the building codes in some eastern countries (i.e. Hong Kong 1995) still use it to evaluate the summer thermal gains for commercial buildings.

2.2.4 Fictitious Ambient Temperature

Nilsson (1994, 1997) developed a simplified analysis tool for the evaluation of a whole building behaviour, based on the preparation of duration diagrams describing the different variables in the thermal zone heat balance equation. However, the principle and the equations are here described since within this tool the dynamic behaviour is introduced by an adjusted reference outdoor temperature for the steady-state calculation of the heat transfer through the building envelope. The principle underneath this approach is that massive constructions are more influenced by the outdoor temperature history than by its instantaneous value.

Equation 2.4 was therefore analytically developed in order to calculate the alternative outdoor temperature, which was called fictitious ambient temperature (FAT). In order to allow easier evaluations when outdoor air temperature values are known at defined time intervals (e.g. hourly steps), Eq. 2.5 was also provided.

$$T^*_{a,t} = T_{a,t} - \left[\left(T_{a,t} - T_{a,0}\right)\mathrm{e}^{-\frac{t}{\tau^*}}\right] \tag{2.4}$$

$$T^*_{a,N} = T_{a,N} - \left[\left(T_{a,N} - T^*_{a,N-1}\right)\mathrm{e}^{-\frac{\Delta t}{\tau^*}}\right] \tag{2.5}$$

where
T^*_a is the fictitious ambient temperature [K]
T_a is the outdoor air temperature [K]
τ^* is the time coefficient [s]
Δt is the time interval [s]
t,N represents the current time
0 represents the initial time

Within the above equations, time coefficient is the characteristic describing the dynamic thermal behaviour of the envelope. Differently from the time constant used in other applications (see also the CEN method in the Chap. 3), this time coefficient is adopted in case of hourly changing outdoor temperature and is calculated considering only the envelope elements (instead of the entire building internal mass) and in their whole thickness.

The envelope heat capacity is described as a lumped mass, meaning that no temperature gradient throughout the mass is taken into account and so the insulation position in the structure cross-section has no effect on the heat transfer: adjustment standard values (ξ) were therefore developed to distinguish between specific layers layouts with the same lumped mass value (Eq. 2.6).

$$\tau^* = \frac{\xi \sum cm}{\sum UA} \tag{2.6}$$

where
ξ is the correction coefficient due to the layers layout
c is the specific heat [J/(kg K)]
m is the mass [kg]
U is the heat transmittance [W/(m^2 K)]
A is the surface [m^2]

Reference values of the correction coefficient were derived and listed in tables for basic layouts (such as massive layers both on the inside and on the outside, only on the inside, only on the outside, and insulation layers both on the inside and on the outside) on the basis of the difference between the fictitious ambient temperature method results and corresponding dynamic simulation results.

2.3 Applications

A larger research carried out in Politecnico di Milano surveyed some of these international simplified methods, for which useful data are available (Ferrari and Zanotto 2010a), and applied them to four different wall solutions, having different aeric mass but the same U-value (and therefore the same steady-state performance), in order to understand their effectiveness (Ferrari and Zanotto 2010b). The results of this research are reported in the present section.

The chosen wall samples are the same considered in the comparison between the finite element calculations and the climatic chamber tests (Chap. 1), and their characteristics are described in Table 2.1:

- a heavyweight masonry wall ("Heavy"), made by one layer of high density hollow bricks;
- a medium/heavyweight wall ("Mid-Ins"), made by outdoor face bricks and a hollow brick layer with insulation within;
- a medium/heavyweight wall ("Ext-Ins") with hollow bricks and exterior insulation;
- a lightweight wall ("Light") composed by an insulation sandwich panel.

The outside boundary condition was set as the heat flow resulting from the standard outdoor surface heat transfer coefficient (EN ISO 6946 2007) and the hourly sol-air temperature for a wall facing South in the city of Rome, calculated according to the Test Reference Year (TRY) climatic data. Similarly, as inside boundary condition the heat flow resulting from the standard indoor surface heat transfer coefficient and a constant temperature (20 °C for the heating season and 26 °C for the cooling season) was used.

Table 2.1 Chosen wall types and related main thermal characteristics

	Heavy	Mid-Ins	Ext-Ins	Light
Thickness [m]	0.480	0.365	0.400	0.125
U-value [W/(m^2 K)]	0.298	0.299	0.306	0.312
Aeric mass [kg/m^2]	431	343	301	35
Heat capacity [kJ/(m^2 K)]	368	291	259	28

2.3.1 M Factor

Starting from Fig. 2.1, the M factor values were derived for the four wall types and for the city of Rome, which is characterised by 1415 heating degree-days. According to these adjustment factors, the corrected U-value were calculates, as shown in Table 2.2.

As design outside temperature for the heating peak load calculation, the lowest value from the Rome TRY data (−0.8 °C) is used. In case of the finite element analysis, the peak load has been derived as the maximum heat flow density rate effect due to the minimum sol-air temperature solicitation. In this way the positive heating effect of the daily solar radiation on the element surface is taken into account.

As shown by Fig. 2.2, the winter peak load prediction according to the M factor and the one according to the finite element analysis method are very different, except for the lightweight construction, which in both cases shows a behaviour consistent with the steady-state approach. According to both methods, the maximum difference can be seen between the lightweight and the heavyweight constructions: this difference is 23 % in case of the M factor method and 46 % in case of the finite element analysis. This discrepancy could be explained, in large extent, by the fact that the M factor was developed in the 1970s, when the envelope average U-values were much higher than the 0.30 W/(m^2 K) of the chosen wall types.

Furthermore, the simplified model is based on a lumped heat capacity approach, bringing to deceiving results when dealing with walls with similar frontal mass but different layouts, for instance in case of the two medium-weight constructions.

2.3.2 CLTD

In order to implement the CLTD method, four wall types were selected from the reference tables provided by ASHRAE (1989), as the most similar constructions (in terms of U-value, frontal mass and layout) to the chosen samples described in Table 2.1. Since, again, the CLTD data were derived for walls with higher average U-values than the studied ones, the correction method for additional insulation was

Table 2.2 Corrected U-values determination according to the nominal U-values and the M factor for the four wall types

	Heavy	Mid-Ins	Ext-Ins	Light
U-value, nominal [W/(m^2 K)]	0.298	0.299	0.306	0.312
M factor	0.81	0.84	0.85	1.00
U-value, corrected [W/(m^2 K)]	0.241	0.251	0.260	0.312

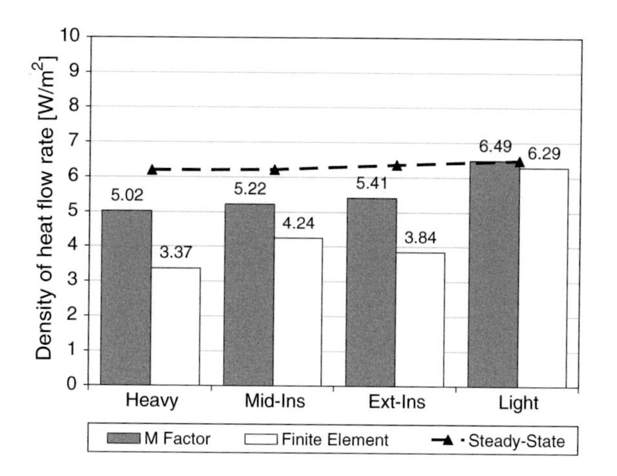

Fig. 2.2 Comparison of the winter peak loads calculated by the means of the M factor, the finite element analysis and the steady-state methods

applied. According to this method, the wall category needs to be upgraded for every 1.23 (m^2 K)/W increase in the thermal resistance value: the first subsequent category has to be adopted unless the insulation is outward, in which case the second subsequent one should be used. If, after this upgrade, the wall is above category A, a temperature difference value is applied for a further category (A^+).

Table 2.3 shows the corrected CLTD values, determined according to the previous consideration and the correction factors in Eq. 2.2. In this study, in particular, an orientation adjustment of 0.5 °C and a surface finishing correction factor of 0.83 (for medium-coloured surfaces) were adopted.

As design outside air temperature for the cooling peak load calculation, the value connected to the highest sol-air temperature from the TRY data (34 °C) is selected, while the design inside air temperature is set to 26 °C, consistently with the Italian common practice.

Figure 2.3 shows the peak density of heat flow rate results for the summer season according to CLTD and finite element analysis. They are quite different, and it can be explained considering the particularly high amount of adjustments necessary to adapt the ASHRAE reference wall types to our four samples. In fact, it was difficult to point out categories matching to our four types, since the selection is based both on the layout descriptions and on the thermal characteristics, and the Italian and U.S. construction traditions are very different.

Table 2.3 Corrected CLTD determination according to the wall types category, increased insulation and the adjustment factors the four wall-types

	Heavy	Mid-Ins	Ext-Ins	Light
Starting category	D	C	B	G
Insulation corrected category	A^+	B	A^+	F
Base CLTD	9.40	11.00	9.40	26.00
Corrected CLTD	12.32	13.65	12.32	26.10

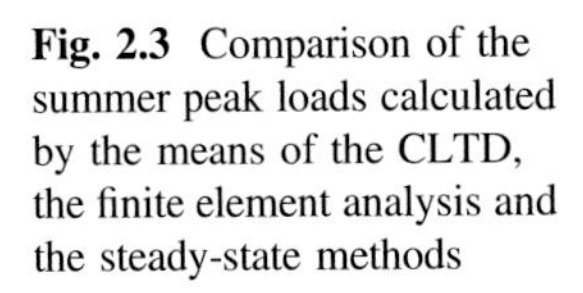

Fig. 2.3 Comparison of the summer peak loads calculated by the means of the CLTD, the finite element analysis and the steady-state methods

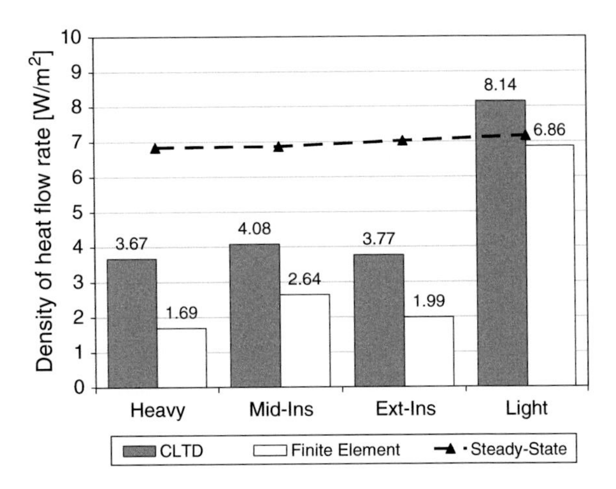

It has to be noted that in this method the layers configuration, taken into account together with the lumped heat capacity, affects the results: the advantage of this approach can be seen in the values regarding the two medium-weight walls, which show the same trend of the finite element analysis ones.

As for the winter peak load, the maximum difference according to both methods is between the lightweight and the heavyweight walls, for 54 % in case of the CLTD method and 75 % in case of the finite element analysis method.

2.3.3 *FAT*

As explained in Sect. 2.2.4, although this method was originally developed in the framework of a whole building behaviour evaluation, the principle and the equations are here applied to isolated element analysis in first approximation.

According to Eq. 2.6 and to the data provided by Nilsson (1997), the characteristic time coefficients for the selected wall types have been derived (Table 2.4).

Later on, the hourly fictitious ambient temperatures for the winter and summer seasons have been calculated according to Eqs. 2.4 and 2.5, and used in the steady-state equation to find the transmission heat exchange. The resulting seasonal heat exchange values have been afterwards compared to the ones from the finite element analysis.

The heating season (November 1st–April 15th) and the cooling season (July 1st–September 15th) have been set according to the Italian praxis.

Figure 2.4 shows the seasonal balance of winter and summer heat exchanges through the selected walls according to the steady-state approach, the FAT method and the finite element analysis.

The values from the simplified method are similar to the ones from the exact solution of the heat conduction equation, as well as to the steady-state results. The

Table 2.4 Time coefficient determination according to the wall types category for the four wall-types

	Heavy	Mid-Ins	Ext-Ins	Light
Category (outwards-inwards)	Massive-massive	Massive-massive	Light-massive	Light-light
Correction factor (ξ)	0.11	0.11	0.08	0.38
Time coefficient [h]	37.84	29.70	18.77	9.54

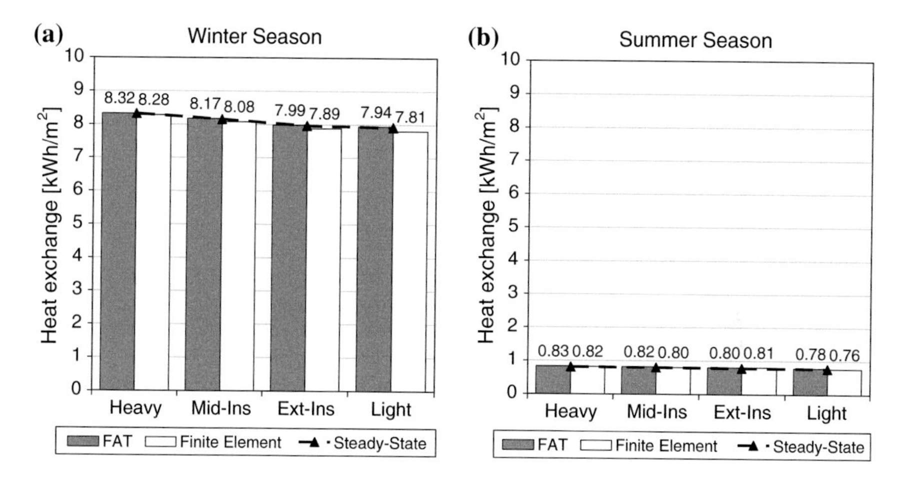

Fig. 2.4 Comparison of the winter and summer seasonal heat exchanges calculated by the means of the fictitious ambient temperature (FAT), the finite element analysis and the steady-state methods

energy balance in the long period, in fact, is similar independently from the magnitude of the heat flow oscillations, which occur around the same average value because of the same environmental conditions. Actually, only considering the whole building performances (including the effects of heat gains and ventilation inside the building), as foreseen by the conventional application of the FAT method, it is possible to appreciate differences even in terms of seasonal balance.

2.3.4 *Lessons Learned*

Regarding the simplified methods to adjust the steady-state heat exchange calculation, the analysed ones aiming at the peak load calculation showed a very small applicability, due to the way they were developed. A possible extension of their use could be foreseen only in case of the development of a wider reference data range (regarding CLTD this has already been done by international studies, e.g. Bansal et al. 2008).

Considering the climatization seasonal need, the study demonstrates that the effect of the thermal mass in different constructions is negligible in the long term (i.e. monthly or seasonal) evaluation of the mere heat exchange through the envelope. In fact it does not consider the local maxima and their interaction with the other unsteady parameters that strongly affects the zone energy balance (i.e. the internal heat sources, the ventilation heat losses, and mass effect due to every other construction element, as described also in Chap. 3). Therefore the adoption of parameters approximating the dynamic behaviour by the means of simplified parameters adopted in the single element analysis will always give inconsistent results.

References

ASHRAE, *Handbook of Fundamentals* (American Society of Heating, Refrigerating and Air-conditioning Engineers, Atlanta, 1972)

ASHRAE, *Handbook of Fundamentals* (American Society of Heating, Refrigerating and Air-conditioning Engineers, Atlanta, 1989)

ASHRAE, *Handbook of Fundamentals* (American Society of Heating, Refrigerating and Air-conditioning Engineers, Atlanta, 2001)

K. Bansal, S. Chowdhury, M.R. Gopal, Development of CLTD values for buildings located in Kolkata. India. Appl. Therm. Eng. **28**(10), 1127–1137 (2008)

C.S. Barnaby, A survey of simplified techniques for calculating energy effects of building mass, in *Proceedings of the Building Thermal Mass Seminar*, Knoxville TN, 2–3 June 1982

EN ISO 6946, *Building components and building elements—Thermal resistance and thermal transmittance—Calculation method* (European Committee of Standardization, Brussels, 2007)

S. Ferrari, V. Zanotto, Simplified indices assessing building envelope's dynamic thermal performance: a survey, in *Proceedings of the CIB World Congress 2010—Building a Better World*, Salford, 10–13 May 2010a

S. Ferrari, V. Zanotto, Assessing building envelope's dynamic thermal performance through simplified parameters, in *Proceedings of PALENC 2010—Cooling the Cities—The absolute priority*, Rhodes, 29 September-1 October 2010b

R.D. Godfrey, K.E. Wilkes, A.G. Lavine, A technical review of the M factor concept, in *Proceedings of the Conference on the Thermal Performance of the Exterior Envelopes of Buildings*, Kissimmee FL, 3–5 December 1979

Hankins and Anderson, *Report on the effect of wall mass on the storage of thermal energy* (Masonry Industry Committee, Boston, 1976)

Hong Kong Building Authority, Code of practice for Overall Thermal Transfer Values in buildings (1995), http://www.bd.gov.hk/english/documents/code/e_ottv.htm. Accessed 21 Dec 2009

Ö. Kaşka, R. Yumrutaş, Experimental investigation for total equivalent temperature difference (TETD) values of building walls and roofs. Energ. Convers. Manag. **50**(11), 2818–2825 (2009)

P.E. Nilsson, *Heating and cooling requirements in commercial buildings—a duration curve model including building dynamics*, Ph.D. Thesis, Chalmers University of Technology, 1994

P.E. Nilsson, Short-term heat transmission calculations by introducing a fictitious ambient temperature. Energ. Build. **25**, 31–39 (1997)

W.J. Van der Meer, Effective U values: a method for calculating the average thermal performance of building components. (New Mexico Energy Institute—University of New Mexico, Albuquerque, 1978)

H.C. Yu, The M-Factor: a new concept in heat transfer calculations. Consult. Eng. **51**(1), 96–98 (1978)

R. Yumrutaş, Ö. Kaşka, E. Yıldırım, Estimation of total equivalent temperature values for multilayer walls and flat roofs by using periodic solution. Build. Environ. **42**(5), 1878–1885 (2007)

Chapter 3
Implications of the Assumptions in Assessing Building Thermal Balance

Abstract The assessment of building energy performance in Europe is widely based on standards that usually refer to the simplified method provided for by EN ISO 13790 (2008). In this simplified method, the estimation of heat transfer is steady-state based, and the building energy balance takes into account the behaviour of the building when interacting with the unsteady parameters by introducing a corrector factor related to the thermal mass effect (i.e. utilization factor). Nevertheless, this method, called "quasi-steady-state", is often unreliable compared to the results provided by a detailed dynamic simulation, in particular concerning the assessment of the thermal need in warmer climates. This chapter highlights this critical point through case studies applications. Additionally, since the common goal of the building energy assessment is to depict a building with an average usage pattern, the implications concerning the assumptions of the main input parameters to be adopted in the building heat balance assessment are also investigated, showing their very high importance in influencing the final climatization need.

Keywords Building energy performance assessment · Simplified building thermal balance · Dynamic building behaviour · Building use parameters · Quasi-steady-state method · Thermal mass effect · Utilization factor of effective mass

3.1 European Standard for Assessing the Building Thermal Behaviour

Standard EN ISO 13790 (2008), which addresses the energy performance of buildings, proposes three different approaches to solve the building heat balance and to calculate the heating and cooling need:

– a simplified seasonal or monthly method (called "quasi-steady-state"), which calculates the climatization need as the steady-state balance of all the thermal

S. Ferrari and V. Zanotto, *Building Energy Performance Assessment in Southern Europe*, PoliMI SpringerBriefs,
DOI 10.1007/978-3-319-24136-4_3

gains and losses, taking into account the dynamic effects by introducing correction factors;
- two hourly methods, including a detailed simulation one that requires the use of tools validated according to what stated within the same standard.

The main difference in adopting the quasi-steady-state method or the detailed simulation one is connected to the calculation time step: with an hourly time step several unsteady parameters directly influencing the building performance can be properly taken into account.

The first among these parameters is the external climatic variability (i.e. air temperature and solar radiation), which is represented by hourly values in case of dynamic simulations and by average monthly (or, in older standards, also seasonal) values in case of the quasi-steady-state method. This difference deeply affects the accuracy of the methods in depicting the external climate and their effect on the building behaviour.

Secondly, the indoor environment variables (i.e. occupation, internal heat sources and ventilation rates) are represented by hourly schedules (often a specific one for each parameter) within the dynamic method, and by monthly average values in case of the quasi-steady-state one. These different approaches affect the accuracy in the depiction of the internal heat sources and, in particular, the simultaneity of the phenomena, influencing their overall effects.

Finally, the greatest difference between the two approaches regards the way the heat storage effect of the building elements mass is considered. The dynamic simulation method takes into account this phenomenon within the calculation of the heat flow through each building element, while the quasi-steady-state one approximate it through a correction parameter, called utilization factor, which is implemented in the calculation at the overall heat balance level.

3.2 Comparison Between Simplified Method and Detailed Simulation

The European Committee for Standardization (CEN), in collaboration with the International Standard Organization (ISO), has addressed the problem of approximating the transient behaviour of the building elements in the climatization need calculation procedure and proposes a correction value to be added to the thermal zone steady-state heat balance equation. Because of this approach, the CEN method is commonly defined “quasi-steady-state”.

The heating need is then determined according to the following Eq. 3.1.

$$Q_H = \left(Q_{env,tr} + Q_{ven}\right) - \eta_{gn}\left(Q_{env,sol} + Q_{\text{int}}\right) \tag{3.1}$$

where

Q_H is the heating need [J]
$Q_{env,tr}$ is the heat transfer from the envelope due to temperature variation [J]
Q_{ven} is the heat transfer due to air mass exchange [J]
$Q_{env,sol}$ is the heat transfer from the envelope due to solar radiation [J]
Q_{int} is the heat production due to internal sources [J]
η_{gn} is the gain utilization factor.

The cooling need, on the other hand, can be calculated according to Eqs. 3.2 and 3.3, which are equivalent.

$$Q_C = \left(Q_{env,sol} + Q_{\text{int}}\right) - \eta_{ls}\left(Q_{env,tr} + Q_{ven}\right) \tag{3.2}$$

$$Q_C = \left(1 - \eta_{gn}\right)\left(Q_{env,tr} + Q_{ven}\right) \tag{3.3}$$

where

Q_C is the cooling need [J]
η_{ls} is the loss utilization factor.

The previously mentioned correction value is the utilization factor, which represents the amount of internal and solar heat gains (for the heating season) or of the transmission and ventilation heat losses (for the cooling season) actually capitalised by the effective mass of the zone. This parameter can be easily calculated and is specific for the characteristics of the analysed building: it depends on the heat gains on heat losses ratio and on the building or zone time constant. The time constant is a measure of the building thermal inertia, and is calculated essentially by the means of the effective heat capacity of both the envelope and the internal elements.

Regarding the calculation of the building effective heat capacity, there are several possible approaches:

- a detailed calculation, solving the transfer matrix of all the elements according to EN ISO 13786 (2007);
- a simplified procedure proposed within EN ISO 13786 (2007), called "effective thickness", that considers the heat capacity of the part of the building elements facing the internal volume;
- simplified standard values suggested within EN ISO 13790 (2008) in case of existing buildings, which are assigned according to building construction.

A comparison between the overall energy need evaluation through both the monthly quasi-steady-state method, applied according to EN ISO 13790 (2008) and considering all the different effective heat capacity calculation possibilities, and the detailed dynamic simulation tool Energy Plus (Crawley et al. 2001), internationally recognized and compliant with ANSI/ASHRAE 140 (2007), has been done with reference to three case-study buildings (Fig. 3.1).

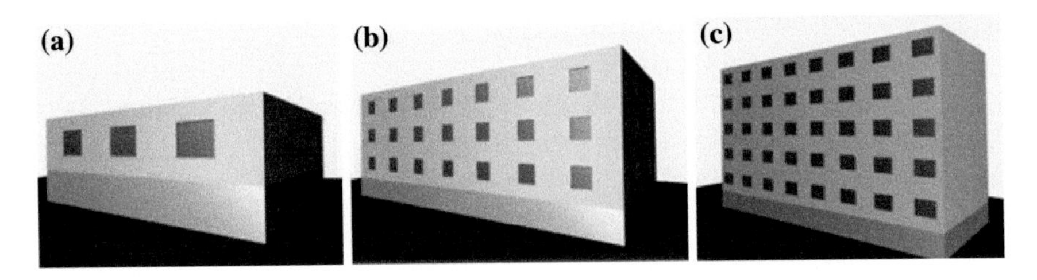

Fig. 3.1 3D schemes representing the case-study buildings. **a** Residential, single-family. **b** Residential, multi-family. **c** Tertiary, open plan office

These buildings are located in Milan, characterized by a relatively cold climate among the Italian ones. The selected buildings are defined by different form factors and final use, as shown in Table 3.1, while the glazed percentage of the façades and the constructions are defined according to the Italian common practice.

Moreover, the buildings are all provided with an unconditioned ground floor (i.e. garage), in order to limit the perturbations on the analysis by neglecting the effects of the heat exchanges with the ground.

To each one of these case-study buildings, three different wall solutions with the same U-value (0.4 W/m^2 K) and different aeric mass were applied:

- a heavyweight wall, made by bricks and insulation (959 kg/m^2);
- a standard wall, made by hollow bricks and insulation (196 kg/m^2);
- a lightweight wall, made by insulation in sandwich (54 kg/m^2).

The desired indoor temperature has been set based on the common praxis, between a winter minimum of 20 °C and a summer maximum of 26 °C.

As previously explained, the main difference between the detailed dynamic method and the simplified monthly/seasonal one concerns the time factor, such as the calculation time schedule and the way the thermal mass effect is taken into account. The time schedules define when and in what extent a certain phenomenon takes place and, regarding buildings are usually referred to occupants presence, equipment or lighting operation, and ventilation operation.

In this study a continuous schedule (24 h/day, 7 days/week) is adopted in case of the quasi-steady-state calculation, commonly allowed as a first simplification in the

Table 3.1 Main characteristics of the case-study buildings

	Residential single-family	Residential multi-family	Tertiary open plan office
Main orientation	North-South	North-South	East-West
Floors	1	3	5
Glazing distribution	Even	Even	Only on main façades
Form-factor [m^{-1}]	0.92	0.44	0.35
Average U-value [W/(m^2 K)]	0.44	0.52	0.54

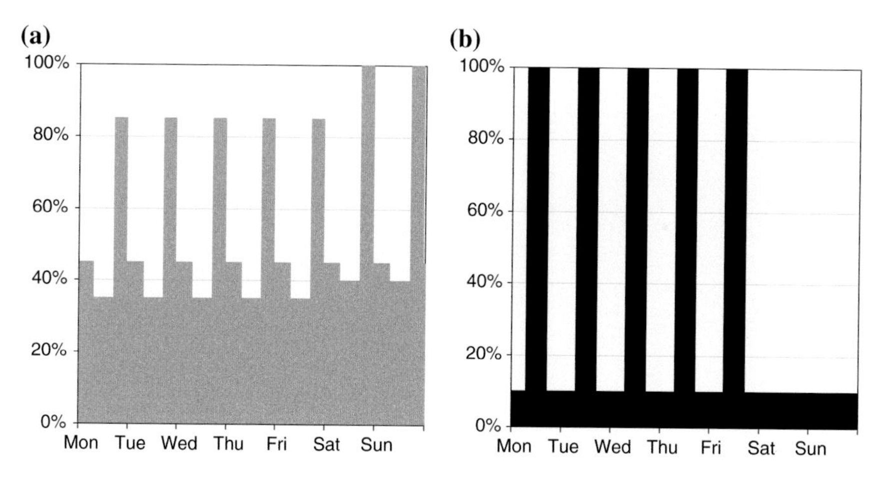

Fig. 3.2 Weekly hourly schedules adopted in the present study for the residential (**a**) and office (**b**) buildings

European standards implementations, while hourly schedules are set in case of the dynamic simulations according to what suggested by EN ISO 13790 (2008), as shown in Fig. 3.2.

Therefore, in order to consistently compare the two calculation methods, the following reference values, considered constant in the quasi-steady-state, have been properly modified in the dynamic simulations in order to obtain the same overall monthly based amounts. The air-change rate, which represents the air volume exchanged between the indoor and the outdoor environments, is set as 0.70 h^{-1}. The internal heat sources (due to occupants presence, equipments and lighting operation)[1] are set according to an overall value, with reference to the floor surfaces, of 4 W/m^2 in case of residential buildings and of 6 W/m^2 in case the office buildings.

Comparing the dynamic and the quasi-steady-state results shown in Figs. 3.3, 3.4 and 3.5, it is evident how the simplified method tends to overestimate the heating need, and to underestimate the cooling need, becoming unreliable in the approximation of the dynamic parameters.

The results also indicate, for both the dynamic and the quasi-steady-state methods, that the change of the only envelope walls construction considered in this study does not deeply affect the thermal inertia effect, since it only concerns the 10 % (office building) and the 16 % (single-family building) of the overall building mass. Only the simplified EN ISO 13790 (2008) approach, which is not based on the actual construction characteristics of the buildings, tends to take into excessive account the mass effect.

[1]It is important to highlight that the quasi-steady state approach and the transient analysis imply very different precision regarding this input parameter, since in case of the dynamic simulation it is possible to adopt separate schedules and values for the various internal sources.

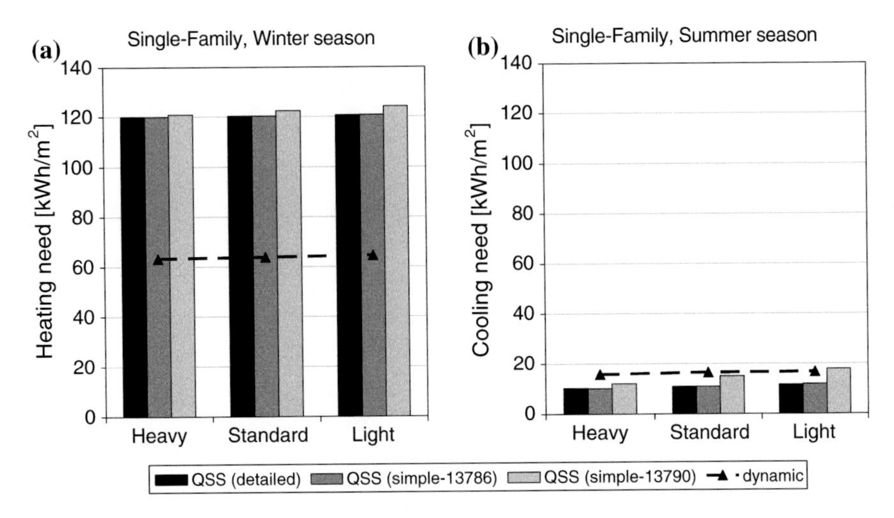

Fig. 3.3 Comparison of the single-family residential building heating (**a**) and cooling (**b**) needs assessed by the means of the quasi-steady-state method, with different approaches in calculating the effective heat capacity, and the ones determined through dynamic simulations

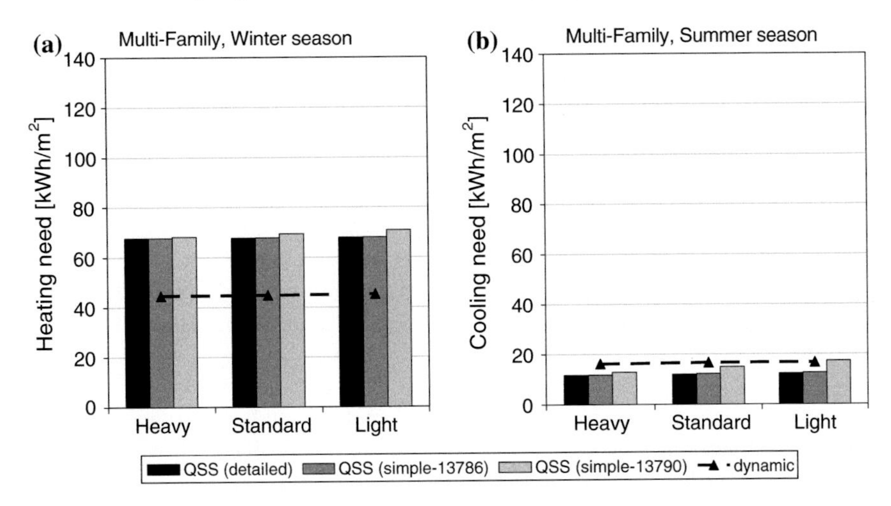

Fig. 3.4 Comparison of the multi-family residential building heating (**a**) and cooling (**b**) needs assessed by the means of the quasi-steady-state method, with different approaches in calculating the effective heat capacity, and the ones determined through dynamic simulations

Finally, it has to be noted that that the gap found in this study between the dynamic and quasi-state-state heating and cooling need results is even attenuated by the different climatic data sources. As shown in Fig. 3.6, the hourly based Test Reference Year file adopted in the Energy Plus simulations (Weather Data for Energy Plus 2012) determines average monthly based values (i.e. air temperature and solar radiation) that are always lower than the ones from the UNI 10349 (1994) used in the quasi-steady-state method.

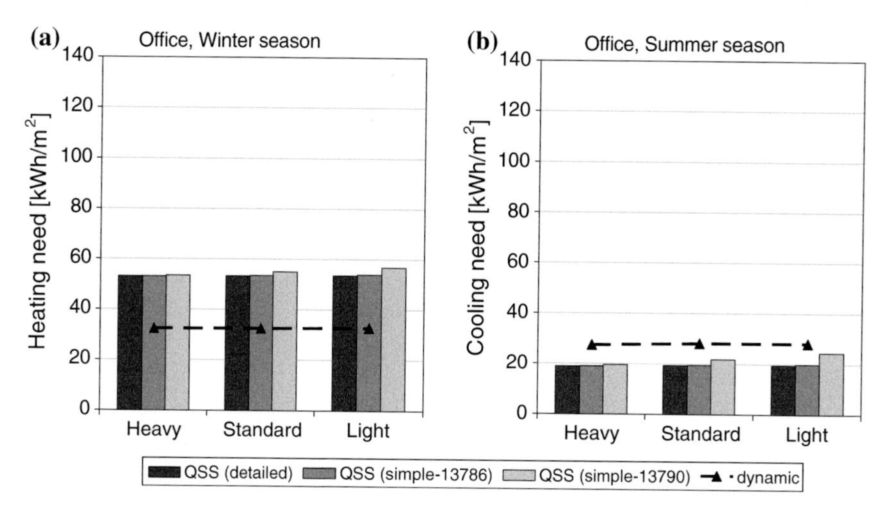

Fig. 3.5 Comparison of the office building heating (**a**) and cooling (**b**) needs assessed by the means of the quasi-steady-state method, with different approaches in calculating the effective heat capacity, and the ones determined through dynamic simulations

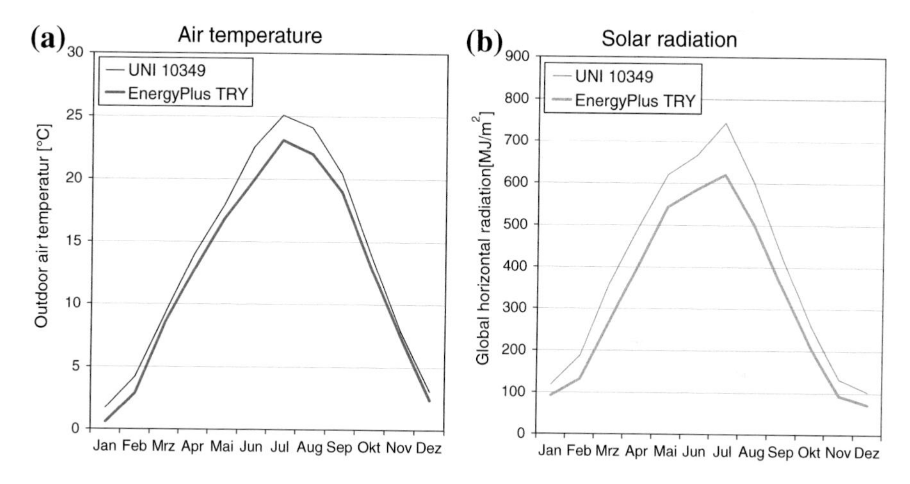

Fig. 3.6 Comparison of the standard reference monthly air temperature and solar radiation values by the Italian norm UNI 10349 and the ones derived by the TRY provided by the U.S. Department of Energy (DOE). City of Milan

Additionally, starting from these basis case-studies, the effects of the other input parameters on the building energy need calculation have been analysed by the means of a sensitivity study, as described in the following section.

3.2.1 *Input Parameters Selection: The Effect on the Thermal Balance*

In the single equations within the main one of the heat balance (see Chap. 1) there are several input parameters that can strongly affect the calculation results: the effects of the different assumptions about these parameters are here enquired with reference to the case study buildings.

3.2.1.1 Set-Point Temperature

The set-point temperature is the indoor desired temperature that determines most of the exchanges between the building zone and the outdoor environment, and is usually set according to local or international standards, such as ISO 7730 (2005) and EN 15251 (2007).

EN 15251 (2007), in particular, proposes temperature limits according to different desired comfort levels, which should be adopted in assessing the indoor environment of different building categories: in case of new or recently retrofitted buildings, for moderate activity uses (such as residential and office occupations) the acceptable temperature limits should be set between a winter minimum of 20 °C and a summer maximum of 26 °C.

In this study, these general average set-point temperatures are used as basic standard values, while further four temperature levels are investigated, ranging around the average ones (Table 3.2).

As shown in Fig. 3.7, the set-point temperature parameter affects the climatization need with a quite high correlation, whether it is calculated by the means of the quasi-steady-state or the dynamic analysis.

3.2.1.2 Air-Change Rate

The air-change rate represents the air volume exchanged between the indoor and the outdoor environments: most local regulations and EN ISO 13790 (2008) allow the adoption of standard reference values according to the building use.

Considering the range of values suggested by several national standards in Europe (Ferrari and Zanotto 2010), the average discharge rate of 0.70 h^{-1} is used as

Table 3.2 Temperature set-point levels investigated in this study, in °C

	Comfort levels				
	I	II	III	IV	V
Heating set-point	22	21	20	19	18
Cooling set-point	24	25	26	27	28

basic reference rate. In this study, four further levels are investigated, regularly ranging between the minimum and maximum values considered by these standards and around the reference rate (Table 3.3).

The results shown in Fig. 3.8 show that, in case of the heating need, the air-change rate settings have almost linear correlations with the energy need. The advantages due to the free-cooling effect of natural ventilation in summer, on the other hand, decrease as the discharge rate increases, tending to a constant cooling need (asymptote) in particular for the dynamic calculation.

In Ferrari and Zanotto (2009) it is also shown how the possible implementation of a night cooling strategy during the summer season would have a greater weight when performing a dynamic analysis, in particular when starting from an already high daily air-change rate. This is due to the fact that it takes into account the lower air temperature during night-time thanks to the hourly weather data, while the reference outdoor temperature in quasi-steady-state case is the monthly average.

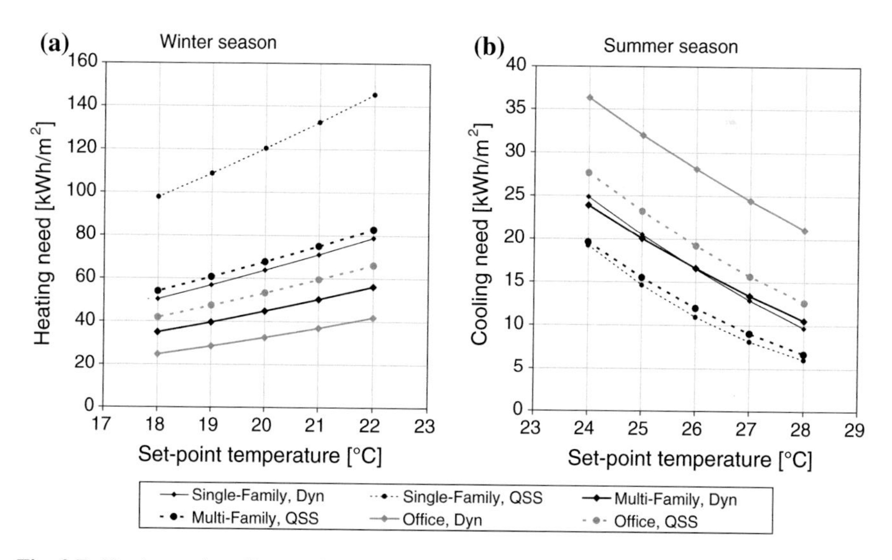

Fig. 3.7 Heating and cooling need variability connected to the considered set-point temperature values

Table 3.3 Air-change rate levels investigated in this study, in h^{-1}

	I	II	III	IV	V
Air-change rate [h^{-1}]	0.3	0.5	0.7	0.9	1.1

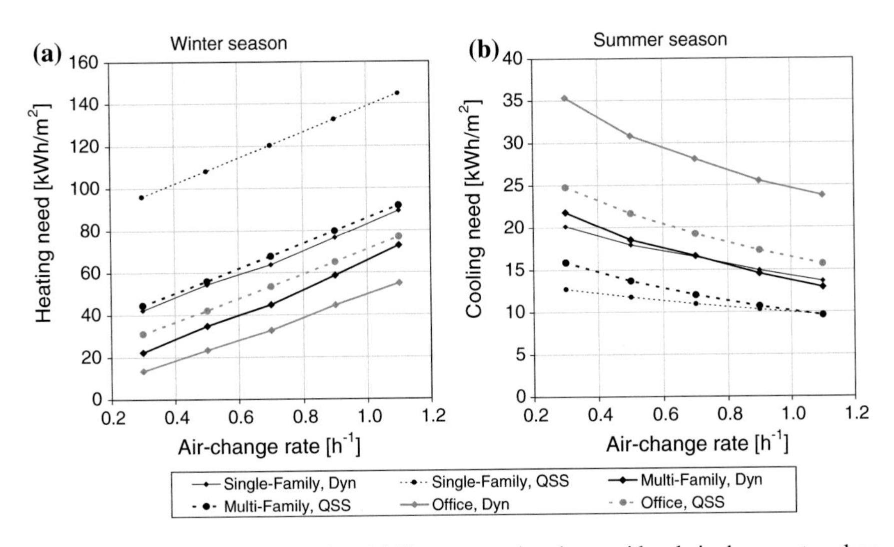

Fig. 3.8 Heating and cooling need variability connected to the considered air change rate values

Table 3.4 Internal heat loads levels investigated in this study, in W/m^2

	I	II	III	IV	V
Residential buildings	2	3	4	5	6
Office building	4	5	6	7	8

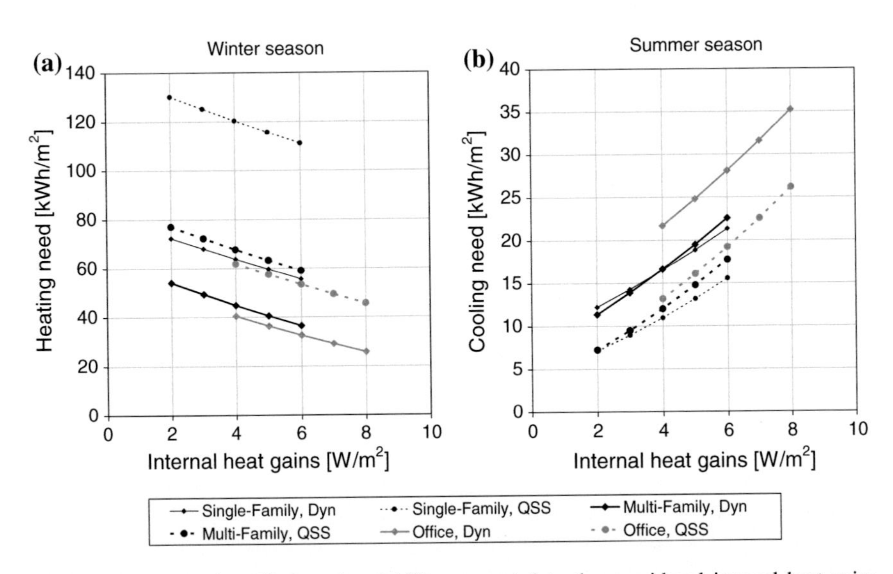

Fig. 3.9 Heating and cooling need variability connected to the considered internal heat gains values

3.2.1.3 Internal Loads

In case of the internal heat sources, too, most of the national procedures prescribe standard reference values according to the building use.

Considering the range of reference data provided by several national standards in Europe (Ferrari and Zanotto 2010) and the ones found in EN ISO 13790 (2008), the average internal heat loads of 4 W/m^2 in case of residential buildings and of 6 W/m^2 in case the office buildings are used as basic reference densities. In this study, four further levels are investigated, regularly ranging between the minimum and maximum values considered in these standards and around the reference rate (Table 3.4).

The results shown in Fig. 3.9 demonstrate, also in this case, an almost linear correlation between the internal heat load input and the climatization need, with a slightly higher dependence in case of the cooling need.

References

ANSI/ASHRAE Standard 140, *Standard method of test for the evaluation of building energy analysis computer program* (American Society of Heating, Refrigerating and Air-conditioning Engineers, Atlanta, 2007)

D.B. Crawley, L.K. Lawrie, F.C. Winkelmann et al., EnergyPlus: creating a new generation building energy simulation program. Energ. Build. **33**, 319–331 (2001)

EN 15251, *Indoor environmental input parameters for design and assessment of energy performance of buildings addressing indoor air quality, thermal environment, lighting and acoustics* (European Committee of Standardization, Brussels, 2007)

EN ISO 13786, *Thermal performance of building components—Dynamic thermal characteristics —Calculation methods* (European Committee of Standardization, Brussels, 2007)

EN ISO 13790, *Energy performance of buildings—Calculation of energy use for space heating and cooling* (European Committee of Standardization, Brussels, 2008)

S. Ferrari, V. Zanotto, EPBD and ventilation requirements: uneven inputs and results in European Countries, in *Proceedings of the 30th AIVC Conference—Trends in High Performance Building and the Role of Ventilation*, Berlin, 1–2 Oct 2009

S. Ferrari, V. Zanotto, EPBD implementation: comparison of different calculation methods among EU Countries, in *Proceedings of PALENC 2010—Cooling the Cities—The absolute priority*, Rhodes, 29 Sept–1 Oct 2010

ISO 7730, *Ergonomics of the thermal environment—Analytical determination and interpretation of thermal comfort using calculation of the PMV and PPD indices and local thermal comfort criteria* (International Standard Organisation, Geneva, 2005)

UNI 10349, *Riscaldamento e raffrescamento degli edifici - Dati climatici* (Ente Nazionale Italiano di Unificazione, Milano, 1994)

Weather Data for Energy Plus (DOE, Energy Efficiency & Renewable Energy, Washington DC, 2012), http://apps1.eere.energy.gov/buildings/energyplus/weatherdata_about.cfm. Accessed 30 Jul 2010

Chapter 4
Thermal Comfort Approaches and Building Performance

Abstract Currently, the suitable indoor temperature is commonly defined according to the thermal comfort theory formulated by Fanger (1970). This approach bases the definition of thermal comfort on mere physics and completely neglects the social and psychological aspects of thermal perception. Moreover, its formulation is completely steady-state, determining a very narrow range of allowable temperatures throughout the year regardless of the outdoor conditions. An alternative approach to defining comfortable temperatures is the adaptive approach, which stems from the results of a wide range of field studies. It assumes that the thermal expectations of the users are linked to the outside climatic conditions on a variable basis. This chapter describes these two approaches and summarises the most relevant adaptive comfort indices. Furthermore, based on the different adaptive formulations, the comfortable temperature trends are derived and compared for two locations in Italy characterized by opposite climatic conditions. Finally, the possible implications on building thermal performance are analysed by means of a case study application.

Keywords Thermal comfort · Adaptive approach indices · Passive cooling strategies · Occupants' control · Indoor comfortable temperatures

4.1 The Standard Approach

Thermal comfort is defined as "that condition of mind that expresses satisfaction with the thermal environment" (ASHRAE 2004), and the way this definition is interpreted can lead to several different physical models.

Modelization depends on the quantitative definition of the quantities representing thermo-hygrometric comfort (Carlucci and Pagliano 2012). There are three different approaches to this problem, corresponding to three different usage patterns:

S. Ferrari and V. Zanotto, *Building Energy Performance Assessment in Southern Europe*, PoliMI SpringerBriefs,
DOI 10.1007/978-3-319-24136-4_4

- psychological model—comfort corresponds to the perception of the thermal environment as satisfactory, which is therefore totally subjective;
- thermo-physiological model—comfort is determined by the smallest possible neural activity (corresponding to the activation of thermal receptors in the skin and in the hypothalamus) reacting to variations in the environmental conditions;
- heat balance model—comfort corresponds to the equilibrium of the heat flows between the environment and the occupant.

The third approach has been widely adopted in thermal standards, given the relatively simplicity of finding quantitative data that can be used in the calculation of a thermal comfort index.

The global diffusion of HVAC (heating, ventilation and air conditioning) systems that could modify the indoor thermal environment resulted in an increased need for a comfort index. Technology has indeed the duty to work towards comfort but, to do so, it has to be programmed through measurable values of precise quantities.

In 1914, Sir L. Hill tried to find a combination of air temperature, mean radiant temperature and air velocity that could correspond to the loss of heat from the human body. Successively, in 1919 ASHRAE formulated the concept of "standard effective temperature" (SET), a function of dry bulb temperature, relative humidity and air velocity, which could be used to describe the effect of these three quantities on the human body. A breakthrough in the definition of these parameters came with the heat balance studies published by Fanger (1970), which led to the definition of a universally recognized comfort index: the predicted mean vote (PMV).

The studies by Fanger had a few important assumptions:

- in the heat balance model, comfort corresponds with the feeling of thermal neutrality;
- the comfort condition is subjective and it is impossible to obtain environmental conditions that satisfy everybody: therefore, intervals of conditions that can be regarded as comfortable have to be defined and a standard number of persons that could still be unsatisfied with the thermal conditions have to be allowed.

The key concept of the model is that the human body, when exposed to constant, moderate thermal conditions and subject to continuous physical activity, tends to produce a heat quantity that balances dissipated heat, without noticeable accumulation. Therefore, the internal production of heat, minus natural losses from evaporation and transpiration (which are regarded as "internal"), is equal to radiative and convective thermal dissipation:

$$H - E_d - E_{sw} - E_r - L = K = R + C \tag{4.1}$$

where

H is the heat produced internally by the human body [W],
E_d is the heat loss caused by the diffusion of water vapor through the skin [W],
E_{sw} is the heat loss caused by the evaporation of sweat from the skin [W],

E_r is the heat loss caused by the latent part of breathing [W],
L is the heat loss caused by the dry part of breathing [W],
K is the heat transfer from the skin to the external body surface (conduction through the clothes) [W],
R is the radiative heat loss from the external body surface [W],
C is the convective heat loss from the external body surface [W].

Fanger realized that taking into account all the quantities involved in the process of heat transfer would make the equation too complicated to be adopted in practice. As a consequence, he developed a simplified version of the model based upon the analysis of the thermoregulation system of the human body.

As a starting point he considered that, in a situation of comfort, the heat transfer between the skin and the external body surface must be null. Consequently, the previous equation can be rewritten as follows:

$$H-(E_d+E_{sw}+E_r+L)-(R+C)=L_o \tag{4.2}$$

where
L_o is the thermal load [W]

The specific quantities taken into account to describe all the component of the previous equation, chosen after a sequence of empirical tests, were the following:

- two values connected to the occupants: activity level and thermal resistance of cloths;
- four values connected to the environment: air temperature, mean radiant temperature, mean air velocity and vapor pressure.

$$\begin{aligned}&\frac{M}{A_{Du}}(1-\eta)-0.35\left[43-0.061\frac{M}{A_{Du}}(1-\eta)-p_a\right]-0.42\left[\frac{M}{A_{Du}}(1-\eta)-50\right]\\&\quad-0.0023\frac{M}{A_{Du}}(44-p_a)-0.0014\frac{M}{A_{Du}}(34-T_a)-3.4\cdot10^{-8}f_{cl}\left(T_{cl}^4-T_{mr}^4\right)\\&\quad+f_{cl}h_c(T_{cl}-T_a)=L\end{aligned}\tag{4.3}$$

where
M is the metabolic activity [W],
A_{Du} is the Du Bois area, quantifying the human body surface [m^2],
$(1-\eta)$ is the inverse of the efficiency concerning the production of heat by the human body, and transforms M/A_{Du} into H/A_{Du}, where H is the heat production,
p_a is the air pressure [Pa],
T_a is the air temperature [°C],
f_{cl} is the insulation factor due to clothing,

T_{cl} is the surface temperature of clothing [°C],
T_{mr} is the mean radiating temperature [°C],
h_c is the convective heat loss coefficient [W/(m^2 K)],
L is the thermal load on the human body per unit area [W/m^2].

In order to reduce the complexity of the formula and transform it into a tool that can be easily applied, Fanger developed a sequence of diagrams that make it possible to compute the value of the different variables.

Even after the creation of these diagrams, the equation still falls short of the original requirements (evaluating the thermal conditions of a room, and individuating the most effective ways to modify them). Assuming the equivalence between comfort and sensation of thermal neutrality, studies were undertaken to build an index of thermal sensation, a quantity that could predict the occupants thermal response, universally valid and based upon the previously mentioned variables.

In order to standardize the results, Fanger used the widely deployed 7-points ASHRAE scale (ASHRAE 2004): from −3, i.e. very cold, to +3, i.e. very warm, with the average value of 0 as the perfect thermal neutrality.

The predicted mean vote (PMV) depends on the thermal load on the human body and on the internal heat production, as shown in the equation

$$Y = \left(0.352 \cdot e^{\frac{M}{A_{Du}}} + 0.032\right) L_o \tag{4.4}$$

where
Y is the predicted mean vote.

The PMV index equation, derived from the comfort one, is as complex as its ‘parent’. As a consequence, Fanger developed a sequence of tables and diagrams that could help calculating an approximated value of the PMV.

Successively, he also tried to calculate the percentage of unsatisfied persons that corresponded statistically to a given PMV: this quantity is important from a normative point of view, since it directly leads to the development of regulations that deal with the issues of comfort. The relation between the PMV and the dissatisfaction level has been developed through the statistical evaluation of thermal votes obtained via an extensive sequence of climatic chamber tests. Considering all the votes outside of the interval between −1 and +1 as unsatisfying, the PPD function (Predicted Percentage of Dissatisfied) shows the correlation between the PMV and the percentage of dissatisfied people.

The main international standards regarding the determination of the indoor environmental conditions (ASHRAE 55 2004; EN 15251 2007; ISO 7730 2005) implement this approach: the Fanger equation is used to determine (according to conventional seasonal clothing level and standard air humidity and velocity) reference temperature ranges related to the building use, which corresponds to defined activity levels.

EN 15251 (2007), for instance, takes into account three requirements categories, and defines ranges of operative temperature, which is a linear combination of air temperature and mean radiant temperature, accordingly:

- category I corresponds to high level of expectation (PPD < 6 %) and is recommended in case of very sensitive and fragile occupants;
- category II corresponds to a normal level of expectation (PPD < 10 %) and is recommended in case of new buildings and renovations;
- category III corresponds to a moderate, yet acceptable, level of expectation (PPD < 15 %), and can be used in case of existing buildings.

The operative temperature limits for office buildings (and all the building uses connected to moderate activity level), for instance, are a minimum 20 °C in winter and a maximum 26 °C in summer.

Fanger's work is of foremost importance, as it resulted in the first universally valid technique for the evaluation of thermal environments. Nevertheless, several critiques were moved towards its formulation.

The tests were run in climatic chamber, in order to closely control the physical variables involved in the equation. Many researchers have criticized this choice, since the tests in a climatic chamber cannot reproduce the conditions of a real environment, which vary much more quickly. In particular, they believed a sequence of real-world tests would have been necessary to validate the study. A lot of real-world tests showed the validity of Fanger's index, but several discrepancies were found in particular when considering extreme climates or transitions between different environments (e.g., moving from an indoor environment to an outdoor environment).

A different observation focuses on the choice of the heat balance model to define comfort. The deterministic approach underlying this model, which reduces thermal sensations to the result of a physiological adjustment, reduces human beings to their biological traits. As a consequence, this model neglects the context and the psychological conditions of the individual, whose influence on the thermal perception isn't considered (Fransson et al. 2007).

A similar point deals with the choice of equating comfort to the neutrality of the heat exchange between the body and the environment (Humphreys and Hancock 2007): the definition of ideal thermal environment as the neutral one is based on a very deterministic approach, which does not take into account the psychological and cultural aspects of comfort and can therefore be questioned. Many studies, in fact, considered the semantic analysis of the experimental questionnaires and detected an influence of the underlying culture on the deep meaning of the used words. This problem is especially valid when referring to outdoor or hybrid environments: an environment can be considered "warm" in the summer without the presence of an actual situation of discomfort. As a consequence, multi-level questionnaires were developed in order to differentiate between 'sensation', 'acceptability' and 'preference', in order to verify the coherence of the different answers.

Lastly, the usage of the ASHRAE scale was criticized due to its "sharpness". Opponents said it could be replaced by methods derived from the field of "fuzzy logic", a technique devised for the purpose of dealing with qualitative evaluations.

In order to bypass the limit of the binary (on/off) approach, fuzzy logic makes it possible to express intermediate positions as a linear combination of the "sharp" values.

4.2 The Adaptive Approach

Because of the previous observations, field studies on environmental conditions and comfort determination inside actual building started taking place from the 1970s, in order to verify the accuracy of the PMV/PPD indices in predicting the actual thermal sensation and preference expressed by the occupants. The results pointed out important differences (Humphreys 1976).

Considering the outcomes of these studies an alternative approach to the definition of comfortable environmental conditions was developed: the adaptive approach. The main assumptions of this approach regard:

- the ability of human beings to adapt themselves to the environmental conditions (through conscious or unconscious changes in their metabolic rate or clothing level) and to interact with the environment in order to adapt it to their needs (through available environmental controls);
- the influence of thermal experience on the occupants' expectations regarding the indoor conditions, which can be short-term, due to the recent weather, or long-term, related to the general climate they are used to.

The huge amount of data collected in these studies (Brager and deDear 1998; Humphreys 1996; Nicol et al. 1995) allowed a statistical analysis which revealed a direct correlation between the indoor comfort temperature and the outdoor one, bringing to the formulation of Eq. 4.5 as the base of the adaptive approach.

$$T_{co} = a \cdot T_{ext,ref} + b \tag{4.5}$$

where

T_{co} is the indoor comfort operative temperature [°C]
$T_{ext,ref}$ is the outdoor reference temperature [°C]
a is the slope of the function
b is the y-intercept of the function [°C]

During the last decades, several specific formulations of the adaptive equation have been developed (Ferrari and Zanotto 2012), according to the results of different studies and to the purpose the equations were formulated for.

Formulations differ both by the equation formulation (a and b values) and by the following specifications.

First of all the outdoor reference temperature determination, which can be the monthly average value, representing the location climate and therefore the long-term expectations, or the running mean one, an average of the mean daily temperatures of the days immediately before the analysed one, weighted according to their time distance, which in this way represents the short-term expectations.

Secondly the acceptability range, indicating the allowable gap from the "ideal" comfort temperature, determined by the means of Eq. 4.5: it can be defined either as constant values consistent to the standard approach or according to a direct correlation with the comfort temperature.

Finally the equation applicability, which usually depends on the season and on the availability of air conditioning or cooling devices.

Among the available adaptive comfort formulations, the most relevant ones are summarized in the following sections.

4.2.1 ASHRAE Equation

The American Society of Heating, Refrigerating and Air Conditioning Engineers (ASHRAE) organized a specific research project on adaptive comfort, managed by deDear and Brager (1997, 2001) and based on a huge amount of statistical data from all over the world. Two equations were developed:

- one for buildings with air-conditioning system, with a lower slope value ($a = 0.11$).
- one for buildings without air-conditioning system, with a higher slope value ($a = 0.31$), since the correlation between comfort and outdoor temperature was found to be stronger than in the previous case.

The second equation was implemented in ASHRAE (2004) in case of summer evaluations (10 °C < $T_{ext,ref}$ < 33.5 °C), by adopting the monthly average as outdoor reference temperature and constant gap values to define 80 % (good comfort level) and 90 % (very good comfort level) acceptability ranges.

4.2.2 ACA Equation

European Union established the research project SCAT (Smart Control and Thermal Comfort), aimed at reducing the energy use due to air conditioning systems by varying the indoor temperature, which resulted in the development of the Adaptive Comfort Algorithm (ACA) equation by McCartney and Nicol (2002) specifically addressing European Countries.

The equation is characterized by a slope value (a) of 0.302 and is valid both for conditioned and unconditioned buildings whenever the outdoor running mean temperature (on a 3.5 days base, which is the reference) is higher than 10 °C; if it is lower, the constant value of 22.8 °C should be assumed (implying the absence of adaptation effects in winter). Concerning the acceptability range, a correlation equation to the comfort temperature value is used, and it is characterised by a negative slope, indicating that the range becomes smaller as the comfort temperature (and therefore the outdoor reference one) increases.

4.2.3 ATG Equation

The Dutch regulation implements the Adaptive Thermal Limit (ATG) standard (Van der Linden et al. 2006), which is based on both the ASHRAE equations, but adds the further comfort range of 65 % (acceptable indoor condition) and takes the running mean temperature (on a 4 days base) as reference.

This standard shows two main peculiarities:

- the adaptive approach is applicable also for winter evaluations (no minimum value for $T_{ext,ref}$ is provided), through the equation characterised by lower slope value (a = 0.11);
- the adaptive approach is applicable in free-running as well as in conditioned buildings, with a distinction in two different categories according to higher or lower adaptation opportunity (i.e. access to environmental controls), using respectively the higher (a = 0.31) or lower (a = 0.11) slope values.

4.2.4 CEN Equation

Partly starting from the SCAT study results, the European Committee for Standardization (CEN) implemented the adaptive approach within the standard EN 15251 (2007), exclusively for summer season analysis (reliability for $10\ °C < T_{ext,ref} < 30\ °C$) in buildings without air-conditioning systems.

The equation is characterized by a slope value (a) of 0.33, and it adopts the running mean temperature (on a 7 days base) as reference and constant values to define 3 different acceptability ranges (consistently with the standard approach classification).

4.3 Approaches Application on a Case Study

In the frame of a wider investigation carried out by the authors Ferrari and Zanotto (2010), the acceptable temperature limits were calculated according to the most restrictive range allowed by the different adaptive formulations. Figures 4.1, 4.2, 4.3 and 4.4 compare the adaptive comfort temperature ranges, for two opposite climatic conditions in Italy (Milano and Palermo), with the “normal level of expectation” of the standard approach (based on the category 20–26 °C).

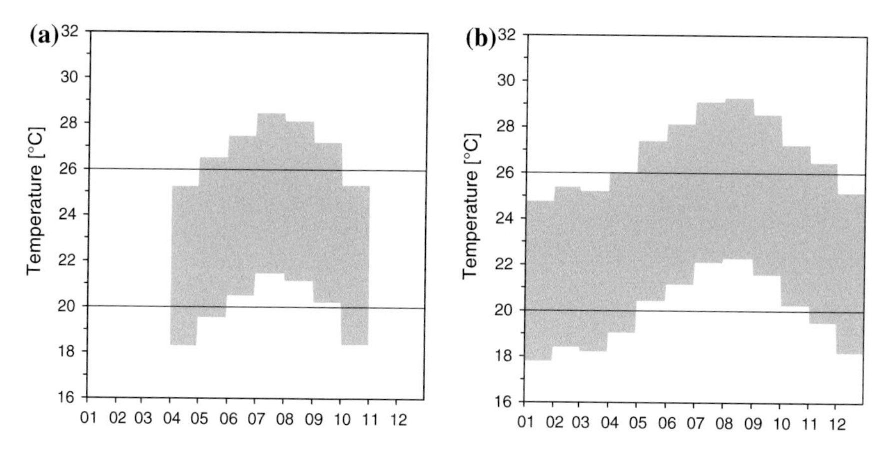

Fig. 4.1 Comparison between the ASHRAE comfort temperature ranges in Milan (**a**) and Palermo (**b**)

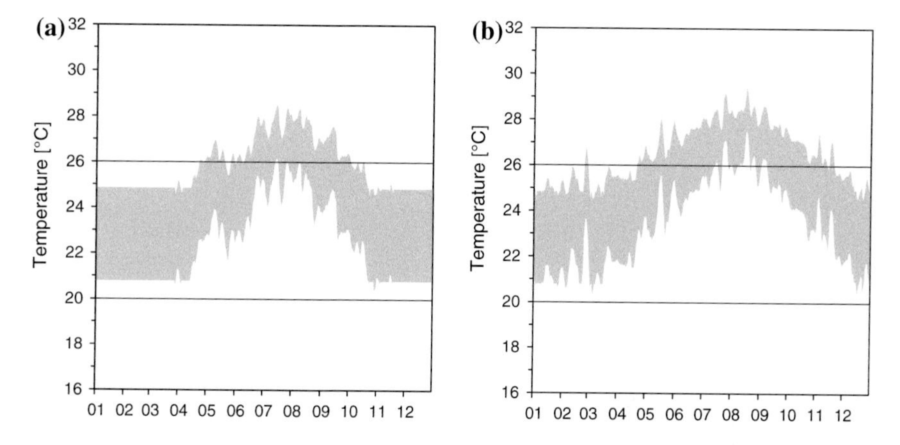

Fig. 4.2 Comparison between the ACA comfort temperature ranges in Milan (**a**) and Palermo (**b**)

The most evident remark from the previous graphs regards the capacity of the adaptive equations to take into account the climate of the building location, whereas the standard practice maintain the same temperature values.

The ACA limit seems to be the less "adaptive" due to the fact that the acceptability range becomes narrower as the outdoor temperature increases (this can be explained considering that the ACA equation was developed considering conditioned buildings, too) and that the slope value is the lowest one (a = 0.302). The CEN range, on the other hand, diverges the most from the standard one, because of the highest slope value (a = 0.33). Interestingly, the ATG results show a slightly more "adaptive" behaviour than the ASHRAE ones, even if they are based on the same equation. This can indicate that the short-term expectations

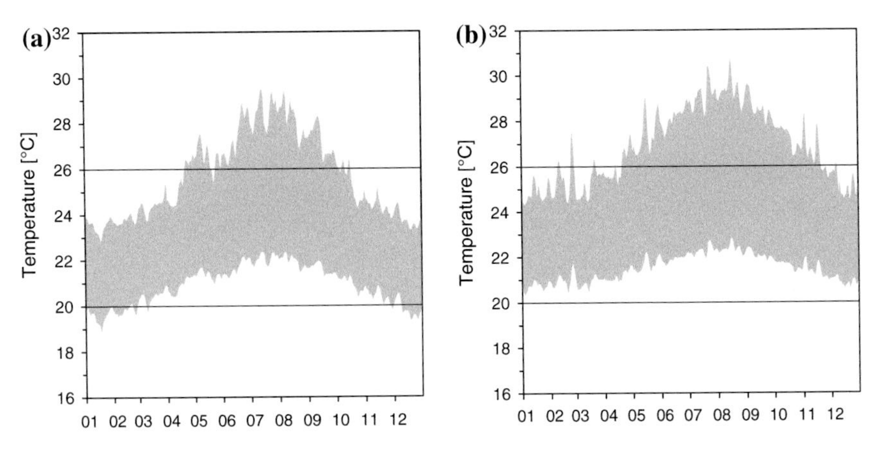

Fig. 4.3 Comparison between the ATG comfort temperature ranges in Milan (**a**) and Palermo (**b**)

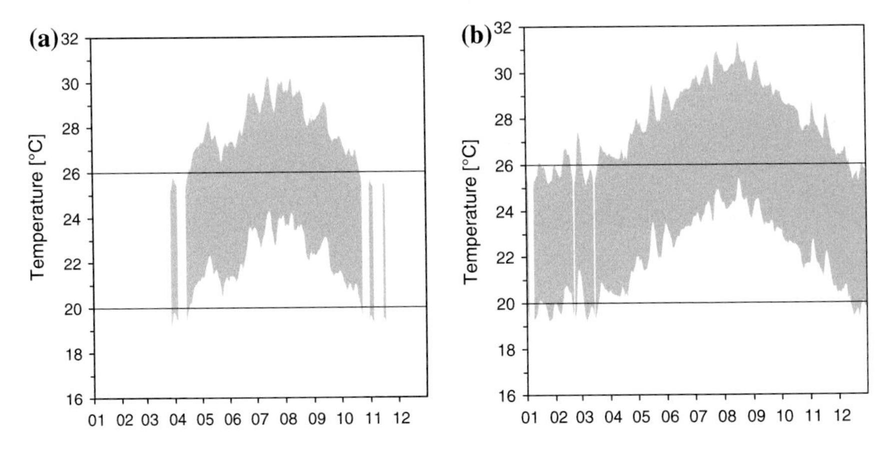

Fig. 4.4 Comparison between the CEN comfort temperature ranges in Milan (**a**) and Palermo (**b**)

(represented by the running mean temperature) can possibly bring to more energy need than the long-term ones (represented by the mean monthly temperature).

Differently from the other adaptive indices, the ATG formulation allows the application of the adaptive approach also for winter evaluations and for conditioned buildings if the equation with a lower slope value ($a = 0.11$) is adopted: moreover, the distinction between buildings provided with adaptive opportunity and the not provided ones is better defined, allowing a broader application of the approach.

For the investigation, the considered case study (an office room naturally ventilated) was simulated without active climatization devices (free-floating conditions), by the means of the dynamic analysis software EnergyPlus (Crawley et al. 2001), in 3 different locations representative of the variability of the Italian climatic conditions (Milan, Rome and Palermo). According to what stated in EN 15251 (2007), the results of the simulation were analysed in terms of discomfort degree-hours (DDH),

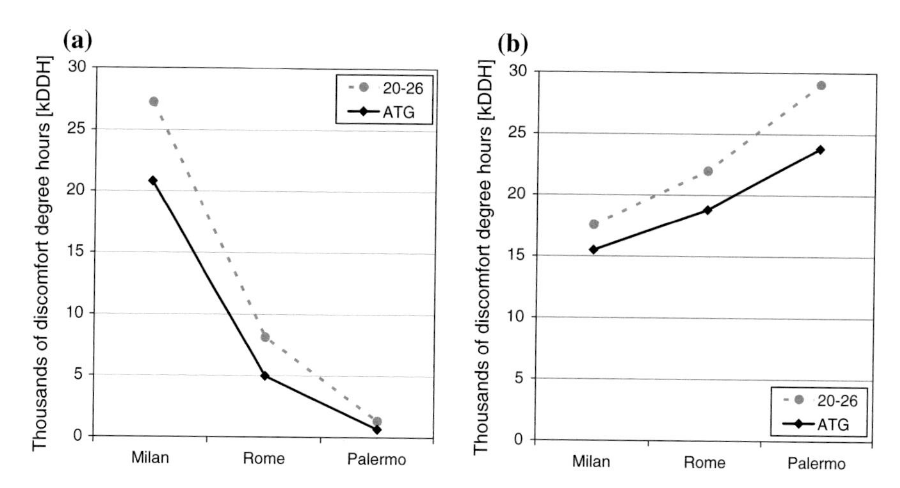

Fig. 4.5 Discomfort degree-hours calculated according to the different indices for the analysed room in the three considered locations (orientation South and base case configuration) for both the winter (**a**) and summer (**b**) seasons

which take into account both the amount of time falling outside the comfort range and the difference between the actual temperature and the desired one: the amount of DDH becomes a general performance index describing the building performance, since it can be considered proportional to the seasonal energy need.

Figure 4.5 reports the results for both the winter and the summer seasons in terms of discomfort degree-hours based on the ATG equation and on the standard approach.

The graphs clearly show that the adaptive approach is more effective in taking into account the climatic peculiarities, since as we move to more extreme climates (regarding both winter and summer) the increase of discomfort degree-hours is evidently less pronounced than in case of the standard approach.

4.4 Building Performance Implications

The adaptive approach is currently implemented in the main international standards (EN 15251 2007; ASHRAE 2004) concerning thermal comfort, but it is usually considered as an assessment method regarding the summer performances of unconditioned buildings, in particular concerning the design phase (through building simulation) and the service phase (through post occupancy evaluation—POE—procedures).

The original separation between naturally ventilated buildings and HVAC buildings, on the other hand, is rather deceiving, since most of office and residential buildings (particularly considering Europe) are not usually part of either two very specific categories. There is a third category, defined as "mixed-mode", which considers all the buildings provided with a hybrid ventilation system, using both operable windows and mechanical systems for air and cooling distribution.

According to Brager and Baker (2008) and to ThermCo (2009), these buildings actually take advantage of the occupants' adaptation capabilities as much as the ones without air-conditioning, in particular due to the importance of the so-called perceived adaptive opportunity in influencing the thermal perception and preference of human beings. It is defined (Baker and Standeven 1996) as a psychological factor affecting the satisfaction of occupants regarding the thermal environment, independently from the actual sensation, due to the availability of environmental controls even when they are not used: this phenomenon does not only concerns thermal comfort, but also health, since the unavailability of controls is also reported as a sick building syndrome cause. The Dutch regulation, based on the ATG equations (Van der Linden et al. 2006), already takes into account this phenomenon, distinguishing buildings on the base of the accessibility of openable windows, thermostats and fans switches.

It is well known that the increase of energy consumptions related to active cooling of buildings is a serious environmental danger, and so it is important to increase the amount of buildings entirely or at least partially relying on passive cooling strategies. From this point of view the adaptive comfort standards are a very important mean (Gossauer and Wagner 2007), which could be adopted in the design phase. By using proper design tools, able to appreciate the related building features (i.e. detailed simulation models), these standards can give a more realistic assessment of the building quality level.

Acknowledging these facts, it is also reasonable to start considering the application of the adaptive approach to the actual management of air conditioning and cooling systems. In this case, there would be two possible and almost subsequent steps of implementation:

- first of all it could be used to control the activation of the cooling system, in order to use it only when the passive means are not enough in attenuating the indoor environmental conditions;
- a further possibility would be to use the adaptive comfort range limits as variable indoor temperature set-points.

In the latter case, it would be very important to harmonize the idea of a "centralized" set-point temperature with the assumption of the approach, which strongly depends on the accessibility of the environmental controls by the occupants. A suitable option to solve any possible conflict consists in the recourse to automated controls and management system, which still can be overridden by manual controls directly activated by occupants (Glicksmann and Taub 1997).

The possible outcomes of basing the HVAC system management on the adaptive ranges can therefore regard:

- the length of the cooling season, which would decrease the related energy consumption;
- the difference between the outdoor and the indoor temperatures, which would not only bring to lower energy consumption due to a lower set-point but would also lead to the installation of more efficient systems/machines.

As an example, the results of the study mentioned in the previous section (Ferrari and Zanotto 2010), showed that the substitution of the standard set-point with the ATG limits could bring to a decrease in the discomfort degree-hours assessment, and therefore of the corresponding cooling needs, between the 10 and the 50 %, depending on the considered Italian location and on the level of implementation of passive cooling strategies.

References

ASHRAE Standard 55, *Thermal Environmental Conditions for Human Occupancy* (American Society of Heating, Refrigerating and Air-conditioning Engineers, Atlanta, 2004)

N. Baker, M. Standeven, A behavioural approach to thermal comfort assessment. Int. J. Sol. Energy **19**, 21–35 (1996)

G.S. Brager, L. Baker, Occupant satisfaction in mixed-mode buildings, in Proceedings of the Air Conditioning and the Low Carbon Cooling Challenge Conference, Windsor, 27–29 July 2008

G.S. Brager, R.J. deDear, Thermal adaptation in the built environment: a literature review. Energy Build. **27**, 83–96 (1998)

S. Carlucci, L. Pagliano, A review of indices for the long-term evaluation of the general thermal comfort conditions in buildings. Energy Build. **53**, 194–205 (2012)

D.B. Crawley, L.K. Lawrie, F.C. Winkelmann et al., Energyplus: creating a new generation building energy simulation program. Energy Build. **33**, 319–331 (2001)

R.J. deDear, G.S. Brager, *ASHRAE RP-884 Final Report—Developibng an Adaptive Model of Thermal Comfort and Preference* (American Society of Heating, Refrigerating and Air-conditioning Engineers, Atlanta, 1997)

R.J. deDear, G.S. Brager, The adaptive model of thermal comfort and energy conservation in the built environment. Int. J. Biometeorol. **45**, 100–108 (2001)

EN 15251, *Indoor Environmental Input Parameters for Design and Assessment of Energy Performance of Buildings Addressing Indoor Air Quality, Thermal Environment, Lighting and Acoustics* (European Committee of Standardization, Brussels, 2007)

P.O. Fanger, *Thermal Comfort: Analysis and Applications on Environmental Technology* (Danisch Technical Press, Copenhagen, 1970)

S. Ferrari, V. Zanotto, Adaptive comfort towards energy savings, in Proceedings of PALENC 2010—Cooling the Cities—The Absolute Priority, Rhodes, 29 Sept–1 Oct 2010

S. Ferrari, V. Zanotto, Adaptive comfort: analysis and application of the main indices. Build. Environ. **49**, 25–32 (2012)

N. Fransson, D. Västfjäll, J. Skoog, In search of the comfortable environment: a comparison of the utility of objective and subjective indicators of indoor comfort. Build. Environ. **45**(2), 1886–1890 (2007)

L.R. Glicksmann, S. Taub, Thermal and behavioural modelling of occupant-comtrolled heating, ventilating and air-conditioning systems. Energy Build. **25**, 243–249 (1997)

E. Gossauer, A. Wagner, Post-occupancy evaluation and thermal comfort: state of the art and new approaches. Adv. Build. Energy Res. **1**, 151–175 (2007)

M.A. Humphreys, Field studies of thermal comfort compared and applied. Build. Serv. Eng. **44**, 5–27 (1976)

M.A. Humphreys, Thermal comfort temperature worldwide—the current position, in Proceedings of WREC, Denver, 15–21 June 1996

M.A. Humphreys, M. Hancock, Do people like to feel "neutral"? Exploring the variation of the desired thermal sensation on the ASHRAE scale. Energy Build. **39**, 867–874 (2007)

ISO 7730, *Ergonomics of the Thermal Environment—Analytical Determination and Interpretation of Thermal Comfort Using Calculation of the PMV and PPD Indices and Local Thermal Comfort Criteria* (International Standard Organisation, Geneva, 2005)

K.J. McCartney, J.F. Nicol, Developing an adaptive algorithm for Europe. Energy Build. **34**, 623–637 (2002)

J.F. Nicol, M.A. Humphreys, O. Sykes et al., *Standards for Thermal Comfort: Indoor Air Temperature Standards for the 21st Century* (Chapman & Hall, London, 1995)

ThermCo, *Interrelation Between Different Comfort Parameters and Their Importance in Occupant Satisfaction. Report for the Thermal Comfort in Buildings with Low-Energy Cooling European Project* (2009)

A.C. Van der Linden, A.C. Boerstra, A.K. Raue et al., Adaptive temperature limits: a new guideline in The Netherlands. A new approach for the assessment of building performance of with respect to thermal indoor climate. Energy Build. **38**, 8–17 (2006)

Chapter 5
Defining Representative Building Energy Models

Abstract The energy need of buildings is strictly related to the local climate and to the construction characteristics (in particular the elements of the building envelope, the ratio between glazed and opaque façade, and orientation). In order to assess the energy performance of different building solutions characterising a large building stock, it can be useful to perform detailed simulation on reference building models. The results obtained can highlight critical issues that are useful for defining adequate policies for improving the built environment. The methodology for characterizing the building energy models, which could constitute a reference for other studies, is described in this chapter as applied in Italy. The set of detailed parameters defining the reference buildings could also be adopted for similar contexts in Southern Europe.

Keywords Building energy simulation parameters · Southern European buildings · Representative building energy models · Building construction ages · Locations representing Italian climatic context

5.1 Definition of the Basis Building Model

Research studies conducted by the authors Ferrari and Zanotto (2011, 2012) were aimed at assessing the thermal behaviour and the energy need of buildings characterising the Italian built environment. In order to perform a generally reliable study, it was necessary to develop building energy models representing typical solutions within the Italian context. These were later simulated in detail (see Chap. 6) by the means of the dynamic software TRNSYS 16 (Klein et al. 2007).

Some parameters (the one described in the present paragraph) are held constant in all models. These are the reference building shape and the building use (i.e. internal heat gains and ventilation rates).

S. Ferrari and V. Zanotto, *Building Energy Performance Assessment in Southern Europe*, PoliMI SpringerBriefs,
DOI 10.1007/978-3-319-24136-4_5

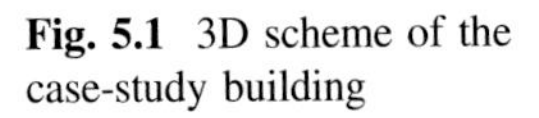

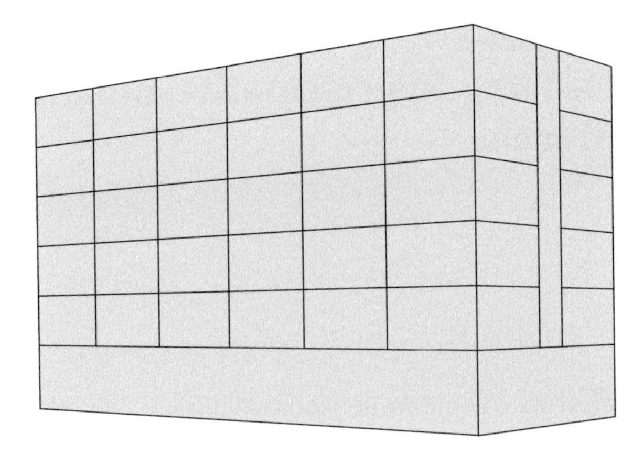

Fig. 5.1 3D scheme of the case-study building

5.1.1 *Building Shape*

The case-study building is characterised by a rectangular plan (30 m × 12 m) and by 6 floors. The first floor is occupied by unconditioned spaces (i.e. parking area or technical rooms) in order to neglect the variability due to the thermal exchanges to the ground, while all the 5 upper floors are divided into a central distribution space (unconditioned) and two side volumes, containing the rooms (6 per floor and side, as can be seen in Fig. 5.1). The rooms and their openings are located on the two main façades allowing to appreciate, through only two rotating options of the model, the performances of the four exposures covering the main orientation cases (North, South, East and West): since the simulation outputs can be required also referring to the building zones having the same exposure, the specific results (e.g. each square meter of floor) can be assumed, even in first analysis, for defining the performance of other buildings having different shapes. Moreover, since these rooms are all identical cellular spaces, nine rooms could be simulated and evaluated in detail, so that all the possible different performances due to the boundary conditions options are covered: the four corner rooms, four edge ones and the central one.

5.1.2 *Internal Heat Loads*

The definition of the internal heat load profiles within the building heat balance calculation (see Chap. 3), is addressed by adopting the suggestions of the official European norm EN ISO 13790 (2008) and implemented within the Italian UNI TS 11300-1 (2008) for more detailed evaluations. These norms define both the absolute value of the heat load (due to occupancy, artificial lighting and equipment) and a basic hourly profile for the occurrence of such loads, as shown in Table 5.1. In this case, an office end-use is assumed in order to precautionary consider highest

Table 5.1 Reference internal heat gains for office buildings according to EN ISO 13790 (2008)

		Offices	
		Office spaces (60 % of the conditions floor area)	Other rooms, lobbies, corridors (40 % of conditioned floor area)
Monday to Friday	07–17	20.0	8.0
	17–23	2.0	1.0
	23–07	2.0	1.0
	Average	9.50	3.92
Saturday to Sunday	07–17	2.0	1.0
	17–23	2.0	1.0
	23–07	2.0	1.0
	Average	2.0	1.0
Average		7.4	3.1

internal loads variability when the cooling energy need of the different solutions will be compared.[1]

The simulation software allows detailing the internal heat loads in disaggregated elements, separating the different sources (occupant, lights and equipment) in order to define their different convective, radiative and latent components. Therefore the global values from the norms are elaborated with weights that were derived from the Switzerland norm SIA 2024 (2006).

In case of the office rooms, the global load is divided in:

- 8.4 W/m^2 due to artificial lighting;
- 5.6 W/m^2 due to electrical equipment;
- 6.0 W/m^2 due to the occupants (i.e. 2 occupants/office and sedentary activity).

These loads are applied according to the time schedules in Table 5.2 and Fig. 5.2.

In case of the central distribution volume, on the other hand, the aggregated values are considered due exclusively to artificial lighting, and therefore a global rate density of 8 W/m^2 was considered and applied according to the time schedule in Table 5.3.

5.1.3 Air Change Rate

The air mass exchange within the building heat balance is considered due to infiltration and simple natural ventilation, which means that the entering air is

[1]It has to be noted that, under the weekly based point of view, the overall amount of these office internal loads, having considered 5 working days, is quite comparable to the overall amount resulting by considering the lower residential heat loads during all the 7 days.

Table 5.2 Hourly schedules related to the internal heat sources of office rooms

	Lighting (%)		Equipment (%)		Occupants (%)	
	Mon–Fri	Sat–Sun	Mon–Fri	Sat–Sun	Mon–Fr	Sat–Sun
00–07	12	12	18	18	0	0
07–17	100		100		100	
17–24	12		18		0	

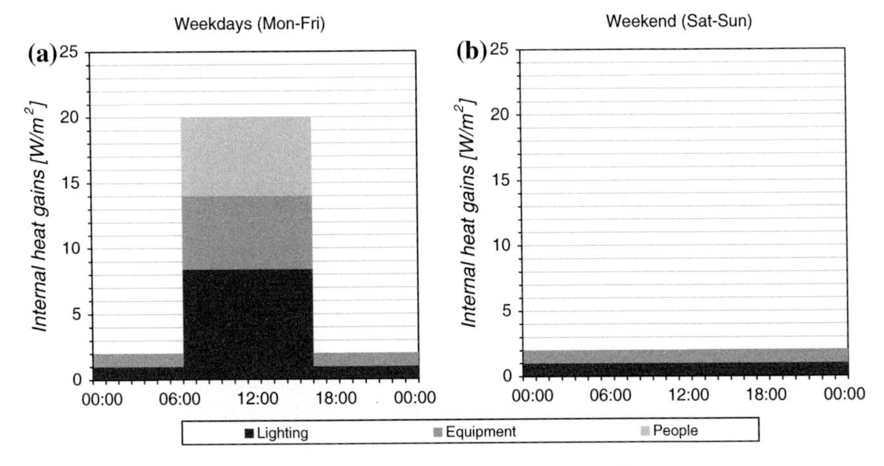

Fig. 5.2 Hourly schedule of the overall internal heat gains, in W/m^2, in the office rooms for the weekdays (**a**) and the weekends (**b**)

Table 5.3 Hourly schedules related to the internal heat source (lighting) of the corridor area

	Lighting (%)	
	Mon–Fri	Sat–Sun
00–07	12.5	12.5
07–17	100	
17–24	12.5	

characterized by the same conditions of the outdoor environment, in order to consider a very common solution, which can be easily adapted and shared also to residential sector.

Regarding the definition of the discharge rate, the reference values are assumed according to norm EN 15251 (2007), which regards the criteria to determine the indoor environment parameters to be considered when assessing the energy performances of buildings. This standard defines the ventilation rate needed to comply with the requirements of indoor air quality and proposes the values in Table 5.4.

In this work the requirements for standard quality (II) in low polluted buildings (this parameter depends on the quality of the construction materials) are used: the global resulting discharge rate is 1.40 l/(m^2 s), corresponding to 1.90 h^{-1}, which should be provided during the occupancy hours. Regarding the time when the

Table 5.4 Discharge rate values in l/(s m^2) for single office rooms according to EN 15251 (2007)

Cat.	For occup. q_p	For building materials q_B			Total q_{tot}			Add when smoking
		Very low pollut.	Low pollut.	Non-low pollut.	Very low pollut.	Low pollut.	Non-low pollut.	
I	1.0	0.5	1.0	2.0	1.5	2.0	3.0	0.7
II	0.7	0.3	0.7	1.4	1.0	1.4	2.1	0.5
III	0.4	0.2	0.4	0.8	0.4	0.8	1.2	0.3

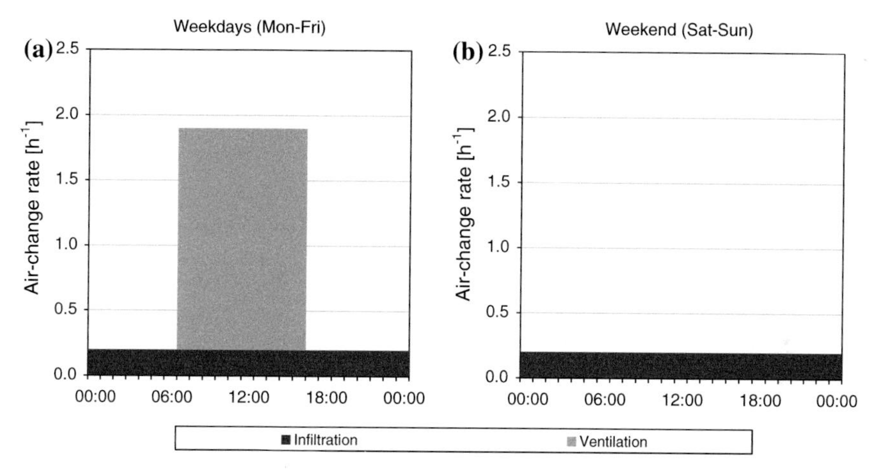

Fig. 5.3 Hourly schedule of the air change rate, in h^{-1}, in the office rooms for the weekdays (**a**) and the weekends (**b**)

building is not occupied, the norm allows taking into account a discharge rate between 0.10 and 0.20 l/(m^2 s) in order to provide a sufficient indoor air quality level at the beginning of the working day: these rates are also coherent with conventional values for natural infiltration through the building envelope. In the simulation model, therefore, it is adopted a constant infiltration rate of 0.15 l/(m^2 s), corresponding to 0.20 h^{-1} (Fig. 5.3).

In the rooms without occupancy (corridors and ground floor) only the natural infiltration rate is considered.

5.2 Definition of the Characterizing Parameters

Some parameters needed by dynamic simulation software are considered variable for defining a matrix of different building models, as described in the following sections: the location—hence climate—and envelope solution, which can also depend on the considered location.

Table 5.5 Heating degree-days (HDD) of the selected locations

	Bolzano	Milano	Trieste	Pescara	Roma	Napoli	Palermo
HDD	2791	2404	2102	1718	1415	1034	751

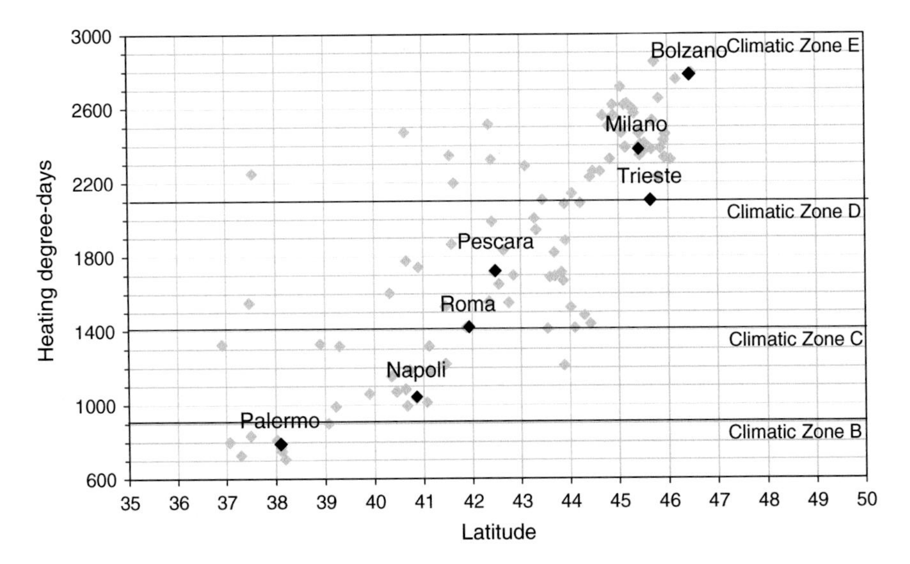

Fig. 5.4 Distribution of the main cities in Italy according to heating degree-days and latitude. The selected locations are highlighted

5.2.1 Building Locations

In order to take into account the wide national climatic variability, the locations are selected according to both their heating degree-days[2] (Table 5.5), which represent the air temperature characteristics and are reported in decree D.P.R. 412 (1993), and, since no official reference cooling degree-days are provided for the Italian cities, to their latitude for distinguishing in first analysis the amounts of solar radiation that influence the building performance in summer.

As a result (Fig. 5.4), seven different locations have been considered to allow following sensitivity analyses of the performances related to climate.

[2]In Italy the heating degree-days values range goes from less than 600, in the national climatic zone A, to over 3000, in the zone F. Nevertheless, buildings are numerically significant in between (zones B–E).

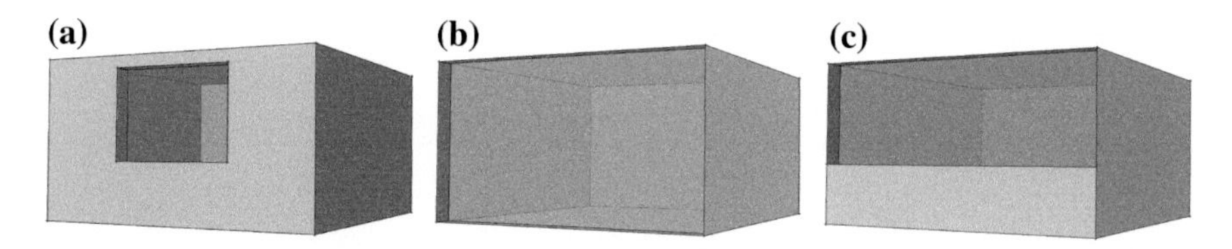

Fig. 5.5 Schemes representing the façade of a single room according to the three considered glazing portion: **a** conventional (new, 1960/80, traditional), **b** completely glazed (new), and **c** largely glazed (1960/80)

5.2.2 Building Constructions

To define the building models set, by limiting the number of the variables, the internal structures remain always the same (based on the widely diffuse ones, hollow bricks walls and concrete and masonry slabs), while the horizontal envelope components (again simple concrete and masonry solutions) only change regarding the presence and thickness of a possible insulation layer, depending on the requirements of the considered construction period.

Hence, the main variation regards the vertical building envelope, and involves a window percentage and a construction solution that are representative of likely practices from three main construction ages: newly built, 1960/80 and very old.

For every construction age, "conventional"[3] vertical envelope solutions have been taken into account, with masonry external walls and a window surface equal to 1/8 of the floor area, which is a common praxis to provide natural ventilation and lighting. Moreover, alternative more glazed and lighter solutions have been considered for the contemporary and 1960/80 ages (Fig. 5.5).

Therefore, the following buildings have been considered:

- two new buildings;
- two buildings from 1960/1980[4];
  - one with conventional façades;
  - one with completely glazed façades;
- one building, which represents the traditional old buildings of the area.
  - one with conventional façades;
  - one with insulated sandwich walls largely glazed;

---

[3]The range of constructions characterising tertiary buildings sometimes differ from the ones adopted in residential buildings. However, all the considered "conventional" envelopes represent widely diffuse solutions and can be adopted also when analysing common residential buildings.

[4]The period between 1960 and 1980 in Italy was the one characterized by the fastest urbanization therefore most of the existing buildings were built during those years (Zuccaro 2002; CRESME 2009).

Table 5.6 U-value limits, in W/(m² K), of the building elements according to D.Lgs. 311 (2006)

	Zone E (BZ, MI, TS)	Zone D (PE, RM)	Zone C (NA)	Zone B (PA)
Vertical opaque walls	0.34	0.36	0.40	0.48
Roofs	0.30	0.32	0.38	0.38
Floors towards unheated rooms	0.33	0.36	0.42	0.49
Elements towards other units	0.80	0.80	0.80	0.80
Windows	2.20	2.40	2.60	3.00
Glasses	1.70	1.90	2.10	2.70

Table 5.7 Aeric mass and periodic thermal transmittance limits according to DPR 59 (2009)

	Aeric mass M_s [kg/m²]	Periodic U-Value Y_{ie} [W/(m² K)]
Vertical walls	230	0.12
Horizontal or inclined surfaces	–	0.20

In particular, regarding the new buildings, the envelope construction was selected in order for them to comply with recent regulations (D.Lgs. 192 2005; D.Lgs. 311 2006; D.P.R. 59 2009) on the elements performances.

First of these limits are the maximum allowable U-values prescribed by Annex C of D.Lgs. 311 (2006) as shown in Table 5.6.

Secondly, the limits of aeric mass or periodic U-value (see definition in Chap. 1) for the vertical and horizontal envelope elements prescribed by D.P.R. 59 (2009) for the towns which do not belong to climatic zone F and where the monthly average value of irradiance on horizontal plane during the month of maximum solar radiation is higher or equal to 290 W/m². Among the locations selected for this work, Pescara, Roma, Napoli e Palermo respect these conditions, but the compliance of these limits, shown in Table 5.7, has been guaranteed also for the other cities.

5.2.2.1 New Conventional

The vertical walls of the new conventional buildings are characterized by two layers of high-density hollow bricks and an insulation layer with thickness variable according to the local U-value requirements (Table 5.8).

The windows cover a surface equal to 1/8 of the office floor area, which is the Italian reference standard to guarantee natural ventilation and lighting and which correspond to the 23 % of the façade (Fig. 5.5). The thermal characteristics of the window elements are defined according to the limits provided by the national regulation (D.Lgs. 311 2006) and are summarised in Table 5.9.

The horizontal envelope components are simple concrete and masonry element with the needed additional insulation to meet the U-value standards. The internal structures remain the same in all locations, characterised by hollow bricks walls and concrete and masonry elements (slabs).

Table 5.8 Detail of the external walls for the new conventional buildings

		Thickness	U-value	Aeric mass	Heat capacity
		[cm]	[W/(m^2 K)]	[kg/m^2]	[kJ/(m^2 K)]
1	Gypsum plaster	1.5	0.33–0.47	~263	~260
2	High density hollow bricks	17.0			
3	Insulation (mineral wool)	7.0–3.5			
4	High density hollow bricks	8.0			
5	Cement plaster	1.5			

Table 5.9 Detail of the transparent envelope elements of the new conventional building

Climatic zone	Glass			Frame	U-value [W/m^{2}K]
	Type	U_{glass} [W/(m^2 K)]	SHGC		
E (MI, BZ, TS)	Low-e double	1.68	0.60	Aluminium	1.77
				Thermal break	
D (PE, RM)	Low-e double	1.87	0.63	Aluminium	2.14
				No thermal break	
C (NA)	Double	2.05	0.63	Aluminium	2.29
				No thermal break	
B (PA)	Double	2.70	0.76	Aluminium	2.81
				No thermal break	

The thermal characteristics of all the considered constructions are summarised in Tables 5.10, 5.11, 5.12 and 5.13.

Globally, the thermal characteristics of the building envelopes are an aeric mass of 284 kg/m^2 and the following U-values:

- 0.47 W/(m^2 K) in climatic zone E;
- 0.53 W/(m^2 K) in climatic zone D;
- 0.59 W/(m^2 K) in climatic zone C;
- 0.69 W/(m^2 K) in climatic zone B.

Table 5.10 Thermal characteristics of the construction elements of the new conventional building for climatic zone E (Milano, Bolzano, Trieste)

	s [cm]	U [W/(m^2 K)]	M_s [kg/m^2]	C [kJ/(m^2 K)]	Y_{ie} [W/(m^2 K)]
Vertical walls	35.00	0.33	264	261	0.06
Roof	53.00	0.30	462	396	0.04
Floor towards unheated parking	38.50	0.32	309	262	0.10
Windows	–	1.77	–	–	–
Internal floors	31.00	0.77	299	257	–
Internal walls	11.00	1.57	92	84	–

Table 5.11 Thermal characteristics of the construction elements of the new conventional building for climatic zone D (Pescara, Roma)

	s [cm]	U [W/(m^2 K)]	M_s [kg/m^2]	C [kJ/(m^2 K)]	Y_{ie} [W/(m^2 K)]
Vertical walls	34.00	0.36	264	260	0.07
Roof	52.00	0.32	462	395	0.04
Floor towards unheated parking	37.50	0.35	308	262	0.11
Windows	–	2.14	–	–	–
Internal floors	31.00	0.77	299	257	–
Internal walls	11.00	1.57	92	84	–

Table 5.12 Thermal characteristics of the construction elements of the new conventional building for climatic zone C (Napoli)

	s [cm]	U [W/(m^2 K)]	M_s [kg/m^2]	C [kJ/(m^2 K)]	Y_{ie} [W/(m^2 K)]
Vertical walls	33.00	0.40	263	260	0.08
Roof	50.00	0.38	461	395	0.05
Floor towards unheated parking	36.00	0.40	308	261	0.13
Windows	–	2.29	–	–	–
Internal floors	31.00	0.77	299	257	–
Internal walls	11.00	1.57	92	84	–

Table 5.13 Thermal characteristics of the construction elements of the new conventional building for climatic zone B (Palermo)

	s [cm]	U [W/(m^2 K)]	M_s [kg/m^2]	C [kJ/(m^2 K)]	Y_{ie} [W/(m^2 K)]
Vertical walls	31.50	0.47	262	259	0.10
Roof	50.00	0.38	461	395	0.05
Floor towards unheated parking	34.50	0.48	307	261	0.16
Windows	–	2.81	–	–	–
Internal floors	31.00	0.77	299	257	–
Internal walls	11.00	1.57	92	84	–

5.2.2.2 New Glazed

The vertical walls of the new conventional buildings are characterized by two layers of hollow bricks and an insulation layer with thickness variable according to the local U-value requirements (Table 5.14).

The windows cover the whole surface of the main façades (Fig. 5.5) and, foreseeing a design which is somehow sensitive to the environmental performance, in order to compensate the unfavourable radiative effect due to the transparency of the element, there have been selected glasses which are more than compliant with the official regulations:

- regarding winter, the U-values are lower than the prescribed ones by 30 %;
- regarding summer, the solar heat gain coefficient is about 0.40 (but with a 0.70 high transmission coefficient for the visible wave-lengths).

The resulting thermal characteristics are summarised in Table 5.15.

The horizontal components (envelope as well as internal) are the same of the new conventional buildings. The internal walls remain the same in all locations, characterised by a dry structure with metal frame and gypsum boards.

Table 5.14 Detail of the external walls for the new glazed buildings

		Thickness [cm]	U-value [W/(m^2 K)]	Aeric mass [kg/m^2]	Heat capacity [kJ/(m^2 K)]
1	Gypsum plaster	1.5	0.33–0.47	~293	~251
2	Hollow bricks	12.0			
3	Insulation (mineral wool)	8.5–5.0			
4	Hollow bricks	12.0			
5	Cement plaster	1.5			

Table 5.15 Detail of the transparent envelope elements of the new glazed building

Climatic zone	Glass			Frame	U-value [W/m^{2}K]
	Type	U_{glass} [W/(m^2 K)]	SHGC		
E (MI, BZ, TS)	Low-e double	1.08	0.39	Aluminium	1.28
				Thermal break	
D (PE, RM)	Low-e double	1.30	0.40	Aluminium	1.46
				Thermal break	
C (NA)	Low-e double	1.40	0.40	Aluminium	1.77
				No thermal break	
B (PA)	Low-e double	1.89	0.41	Aluminium	2.16
				No thermal break	

The thermal characteristics of all the considered constructions are summarised in Tables 5.16, 5.17, 5.18 and 5.19.

Globally, the thermal characteristics of the building envelopes are an aeric mass of 201 kg/m^2 and the following U-values:

- 0.74 W/(m^2 K) in climatic zone E;
- 0.83 W/(m^2 K) in climatic zone D;
- 0.99 W/(m^2 K) in climatic zone C;
- 1.19 W/(m^2 K) in climatic zone B.

Table 5.16 Thermal characteristics of the construction elements of the new glazed building for climatic zone E (Milano, Bolzano, Trieste)

	s [cm]	U [W/(m^2 K)]	M_s [kg/m^2]	C [kJ/(m^2 K)]	Y_{ie} [W/(m^2 K)]
Vertical walls	35.50	0.33	294	252	0.09
Roof	53.00	0.30	462	396	0.04
Floor towards unheated parking	38.50	0.32	309	262	0.10
Windows	–	1.28	–	–	–
Internal floors	31.00	0.77	299	257	–
Internal walls	15.00	1.54	45	49	–

Table 5.17 Thermal characteristics of the construction elements of the new glazed building for climatic zone D (Pescara, Roma)

	s [cm]	U [W/(m^2 K)]	M_s [kg/m^2]	C [kJ/(m^2 K)]	Y_{ie} [W/(m^2 K)]
Vertical walls	34.50	0.36	293	251	0.10
Roof	52.00	0.32	462	395	0.04
Floor towards unheated parking	37.50	0.35	308	262	0.11
Windows	–	1.46	–	–	–
Internal floors	31.00	0.77	299	257	–
Internal walls	15.00	1.54	45	49	–

Table 5.18 Thermal characteristics of the construction elements of the new glazed building for climatic zone C (Napoli)

	s [cm]	U [W/(m^2 K)]	M_s [kg/m^2]	C [kJ/(m^2 K)]	Y_{ie} [W/(m^2 K)]
Vertical walls	33.50	0.40	293	251	0.12
Roof	50.00	0.38	461	395	0.05
Floor towards unheated parking	36.00	0.40	308	261	0.13
Windows	–	1.77	–	–	–
Internal floors	31.00	0.77	299	257	–
Internal walls	11.00	1.57	92	84	–

Table 5.19 Thermal characteristics of the construction elements of the new glazed building for climatic zone B (Palermo)

	s [cm]	U [W/(m^2 K)]	M_s [kg/m^2]	C [kJ/(m^2 K)]	Y_{ie} [W/(m^2 K)]
Vertical walls	32.00	0.47	292	259	0.14
Roof	50.00	0.38	461	395	0.05
Floor towards unheated parking	34.50	0.48	307	261	0.16
Windows	–	2.16	–	–	–
Internal floors	31.00	0.77	299	257	–
Internal walls	11.00	1.54	45	49	–

5.2.2.3 60/80 Conventional

The vertical walls of the conventional building from 1960/80 are derived from CNR (1982) and are made by two layers of hollow bricks with an air layer within (Table 5.20).

Also in this case the windows cover a surface equal to 1/8 of the office floor area (Fig. 5.5).

According to recent studies on the office buildings in Italy (CRESME 2009), more than 50 % of the office buildings from earlier than 1970 have been undertaken retrofit measures, with a wide presence of double glazing in existing offices. Foreseeing a progressive completion of this retrofit campaign, also to comply with the recent acoustic performance requirements, the windows described in Table 5.21 can be considered for this building category.

Table 5.20 Detail of the external walls for the conventional building from 1960/80

		Thickness [cm]	U-value [W/(m^2 K)]	Aeric mass [kg/m^2]	Heat capacity [kJ/(m^2 K)]
1	Gypsum plaster	2.0	0.98	305	263
2	Hollow bricks	12.0			
3	Air	7.0			
4	Hollow bricks	12.0			
5	Cement plaster	2.0			

Table 5.21 Detail of the transparent envelope elements of the 60/80 conventional building

Climatic zone	Glass			Frame	U-value [W/m^2 K]
	Type	U_{glass} [W/(m^2 K)]	SHGC		
All	Double	2.83	0.76	Aluminium	2.91
				No thermal break	

Table 5.22 Thermal characteristics of the construction elements of the conventional building from 1960/80

	s [cm]	U [W/(m^2 K)]	M_s [kg/m^2]	C [kJ/(m^2 K)]
Vertical walls	35.00	0.98	305	264
Roof	36.00	0.91	317	302
Floor towards unheated parking	28.50	1.66	303	258
Windows	–	2.91	–	–
Internal floors	28.50	1.63	298	256
Internal walls	11.00	1.57	92	84

The horizontal components are simple uninsulated concrete and masonry elements. The internal walls are characterised by hollow bricks structures.

The thermal characteristics of all the considered constructions are summarised in Table 5.22.

Globally, the thermal characteristics of the building envelope are a U-value of 1.30 W/(m^2 K) and an aeric mass of 276 kg/m^2.

5.2.2.4 60/80 Sandwich Largely Glazed

The vertical walls of the second building from 1960/80 are derived from CNR (1982) and are made by a sandwich structure with a 10 cm insulation (Table 5.23).

The window surfaces cover the whole main façades except for an opaque sill of 1.10 m, with a resulting glazed percentage for the façade of 63 % (Fig. 5.5).

As for the other 60/80 building, the windows described in Table 5.24 are considered for this building category (CRESME 2009).

The horizontal components are simple uninsulated concrete and masonry elements. The internal walls are characterised by a dry structure with metal frame and gypsum boards.

Table 5.23 Detail of the external walls for the largely glazed sandwich building from 1960/80

		Thickness [cm]	U-value [W/(m^2 K)]	Aeric mass [kg/m^2]	Heat capacity [kJ/(m^2 K)]
1	Gypsum board	2.5	0.36	56	53
2	Insulation (mineral wool)	10.0			
3	Fiber cement board	1.5			

Table 5.24 Detail of the transparent envelope elements of the 60/80 largely glazed building

Climatic zone	Glass			Frame	U-value [W/m^2 K]
	Type	U_{glass} [W/(m^2 K)]	SHGC		
All	Double	2.83	0.76	Aluminium No thermal break	2.91

Table 5.25 Thermal characteristics of the construction elements of the sandwich building from 1960/80

	s [cm]	U [W/(m^2 K)]	M_s [kg/m^2]	C [kJ/(m^2 K)]
Vertical walls	14.00	0.36	56	53
Roof	36.00	0.91	317	302
Floor towards unheated parking	28.50	1.66	303	258
Windows	–	2.91	–	–
Internal floors	28.50	1.63	298	256
Internal walls	15.00	1.54	45	49

The thermal characteristics of all the considered constructions are summarised in Table 5.25.

Globally, the thermal characteristics of the building envelope are a U-value of 1.42 W/(m^2 K) and an aeric mass of 139 kg/m^2.

5.2.2.5 Traditional

The vertical wall construction for the historical buildings is different according to location and related construction traditions. In Italy, in general, buildings from earlier than the second world war have masonry walls either made by bricks or by some kind of stone, which is easily found in the surrounding areas (Zuccaro 2002):

- in the Northern (Milano and Bolzano) and Adriatic (Trieste and Pescara) regions the most used construction involved full bricks;
- in the Southern regions, and in particular in Campania (Napoli) and Sicilia (Palermo), tuff was extremely common.

Regarding the city of Rome, due to its location and importance, a very wide range of constructions were used in time, but the statistically most common involved the use of both full bricks and tuff: in this work a simple tuff wall is used, since in the other location of the same climatic zone (Pescara) full bricks are adopted.

In this way, two main old constructions are considered (Table 5.26), and their adoption depends on the latitude of the building location.

Table 5.26 Detail of the external walls for the traditional buildings

		Thickness [cm]	U-value [W/(m^2 K)]	Aeric mass [kg/m^2]	Heat capacity [kJ/(m^2 K)]
1	Gypsum plaster	2.0	1.08 (MI–PE) 0.98 (RM–PA)	965 (MI–PE) 815 (RM–PA)	818 (MI–PE) 587 (RM–PA)
2	Full Bricks (MI–PE) Tuff (RM–PA)	50.0			
3	Cement plaster	2.0			

In this case, again, the windows cover a surface equal to 1/8 of the office floor area (Fig. 5.5).

As for the 60/80 buildings, the windows described in Table 5.27 are considered for this building category (CRESME 2009).

The horizontal components are simple uninsulated concrete and masonry elements. The internal walls are characterised by hollow bricks structures.

The thermal characteristics of all the considered constructions are summarised in Tables 5.28 and 5.29.

Globally, the thermal characteristics of the building envelopes are:

- U-value 1.33 W/(m^2 K) and aeric mass 612 kg/m^2 for the brick construction;
- U-value 1.28 W/(m^2 K) and aeric mass 536 kg/m^2 for the tuff construction.

Table 5.27 Detail of the transparent envelope elements of the traditional building

Climatic zone	Glass			Frame	U-value [W/m^2 K]
	Type	U_{glass} [W/(m^2 K)]	SHGC		
All	Double	2.83	0.76	Wood	2.91

Table 5.28 Thermal characteristics of the construction elements of the traditional building in Milano, Bolzano, Trieste and Pescara

	s [cm]	U [W/(m^2 K)]	M_s [kg/m^2]	C [kJ/(m^2 K)]
Vertical walls	54.00	1.08	965	818
Roof	36.00	0.91	317	302
Floor towards unheated parking	28.50	1.66	303	258
Windows	–	2.73	–	–
Internal floors	28.50	1.63	298	256
Internal walls	11.00	1.57	92	84

Table 5.29 Thermal characteristics of the construction elements of the traditional building in Roma, Napoli and Palermo

	s [cm]	U [W/(m^2 K)]	M_s [kg/m^2]	C [kJ/(m^2 K)]
Vertical walls	54.00	1.08	965	818
Roof	36.00	0.91	317	302
Floor towards unheated parking	28.50	1.66	303	258
Windows	–	2.73	–	–
Internal floors	28.50	1.63	298	256
Internal walls	11.00	1.57	92	84

References

CNR, *Appendice 2 alla guida al controllo energetico della progettazione - Repertorio delle caratteristiche termofisiche dei component edilizi opachi e trasparenti* (Consiglio Nazionale delle Ricerche, Roma, 1982)

CRESME, *Determinazione dei fabbisogni di e dei consumi energetici dei sistemi edificio-impianto - Caratterizzazione del parco immobiliare ad uso ufficio* (Ente Nazionale per le Nuove Tecnologie, l'Energia e l'Ambiente, Roma, 2009)

D.Lgs. 192, *Attuazione della direttiva 2002/91/CE relativa al rendimento energetico nell'edilizia* (Rome, 2005)

D.Lgs. 311, *Disposizioni correttive ed integrative al decreto legislativo 19 agosto 2005, n. 192, recante attuazione della direttiva 2002/91/CE, relativa al rendimento energetico nell'edilizia* (Rome, 2006)

D.P.R. 59, *Regolamento di attuazione dell'articolo 4, comma 1, lettere a) e b), del decreto legislative 19 agosto 2005, n. 192, concernente attuazione della direttiva 2002/91/CE sul rendimento energetico in edilizia* (Rome, 2009)

D.P.R. 412, *Regolamento recante norme per la progettazione, l'installazione e la manutenzione degli impianti termici degli edifici, ai fini del contenimento dei consumi di energia, in attuazione dell'art. 4, comma 4 della legge 9 gennaio 1991, n.10. (aggiornata dal D.P. R.551/99)* (Rome, 1993)

EN 15251, *Indoor environmental input parameters for design and assessment of energy performance of buildings addressing indoor air quality, thermal environment, lighting and acoustics* (European Committee of Standardization, Brussels, 2007)

EN ISO 13790, *Energy performance of buildings—Calculation of energy use for space heating and cooling* (European Committee of Standardization, Brussels, 2008)

S. Ferrari, V. Zanotto, Energy performance of different office building envelopes within the Italian context, in *Proceedings of the 48th AICARR International Conference—Energy Refurbishment of Existing Buildings—Which Solutions for an Integrated System: Envelope, Plant, Controls*, Baveno, 22–23 Sept 2011

S. Ferrari, V. Zanotto, Office buildings cooling need in the Italian climatic context: assessing the performances of typical envelopes. Energy Procedia **30**, 1099–1109 (2012)

S.A. Klein, W.A. Beckman, J.W. Mitchell et al., *TRNSYS—A Transient System Simulation Program User Manual* (The solar energy Laboratory—University of Wisconsin, Madison, 2007)

SIA 2024, *Standard-Nutzungsbedingungen für die Energie- und Gebäudetechnik* (Schweizerischer Ingenieur- und Architektenverein, Zürich, 2006)

UNI TS 11300-1, *Determinazione del fabbisogno di energia termica dell'edificio per la climatizzazione estiva e invernale* (Ente Nazionale Italiano di Unificazione, Milano, 2008)

G. Zuccaro, *Modello di caratterizzazione tipologica a scala nazionale. Relazione Finale* (2002)

Chapter 6
Energy Performance Analysis of Typical Buildings

Abstract The study reported in this chapter addresses the assessment of the thermal behaviour of typical buildings within the Italian context, taken as representative of the countries in southern Europe. Five representative case-studies are reported for three principal construction ages (newly built, 1960/80, very old), the two more recent periods having seen the use of more glazed and lighter walls as an alternative to more conventional solutions. The performances of the buildings are predicted, through detailed dynamic simulations, with reference to the resulting indoor operative temperatures in order to consider the actual overall thermal comfort sensation provided by the different envelope solutions. Considering that southern Europe is mostly characterized by warm climates, the study also analyses the effectiveness of two basic passive cooling strategies (i.e. shading and night ventilation) in relation to the different building characteristics. Moreover, the implications in the seasonal cooling need assessment when considering the adoption of climate-related set-point temperatures (i.e. adaptive comfort approach), beyond the assumed common standard, are also evaluated.

Keywords Building energy performance analysis · Building construction ages · Southern European building performances · Indoor operative temperature requirement · Passive cooling strategies · Adaptive set-point

6.1 The Set of the Simulations

This study assesses, by the means of dynamic simulation performed with TRNSYS 16 (Klein et al. 2007), the behaviour of typical buildings within the Italian context, which is characterized by warm climates similarly to other southern European countries. The considered building use is for offices, in order to precautionary take into account a higher internal heat loads variability when comparing the cooling energy need of the different solutions. The overall heat amount, weekly based, however, is also comparable to the residential use.

S. Ferrari and V. Zanotto, *Building Energy Performance Assessment in Southern Europe*, PoliMI SpringerBriefs,
DOI 10.1007/978-3-319-24136-4_6

The reference building shape is a sample of five storey building with rectangular plan (30 m × 12 m), and the case-studies are defined based on both different envelope characteristics and window percentage, to be representative of likely practices from three main construction ages: newly built, 1960/80 and very old. The internal structures remain always the same (hollow bricks walls and concrete and masonry slabs), while the horizontal envelope components (again simple concrete and masonry solutions) only change regarding the presence and thickness of a possible insulation layer, which depends on the reference U-value standards for the different considered locations.

For this study seven locations are chosen in order to represent the wide range of the Italian climatic conditions, both according to different national heating degree-days, indicating the variety of winter characteristics in terms of air temperature, and to different latitudes, distinguishing in first analysis the amounts of solar radiation that influence building performance predominantly in summer.

For every construction age, "conventional"[7] vertical envelope solutions, which can also refer to residential building, have been taken into account, with masonry external walls and a window surface equal to 1/8 of the floor area, based on the diffuse Italian praxis to provide natural ventilation and lighting. Moreover, alternative more glazed and lighter solutions have been considered for the contemporary and 1960/80 ages.

The rooms and their openings are located on the two main façades in order to appreciate, through only two rotating options of the model, the performances of the four exposures covering the main orientation cases (North, South, East and West).

Hence, the representative building simulation models are defined based on five different envelope solutions (Tables 6.1), seven climatic locations and four main orientations (detailed description in Chap. 5).

Furthermore, the analyses of buildings energy performance in Italian climatic context have to carefully take into account the summer behaviour: the matrix of the

Table 6.1 Thermal characteristics of the overall envelope for the different solutions

		New conventional	New glazed	1960/80 conventional	1960/80 sandwich largely glazed	Traditional
U-value [W/ (m^2 K)]	MI	0.47	0.74	1.30	1.42	1.33
	BZ					
	TS					
	PE	0.53	0.83			
	RM					1.28
	NA	0.59	0.99			
	PA	0.69	1.19			
Areic mass [kg/m^2]		~284	~201	276	139	612/536
Heat capacity [kJ/ (m^2 K)]		~260	~171	243	127	~530

building simulation models is here implemented, by considering the effects of two basic passive cooling strategies (shading and night ventilation) and by setting, as an alternative to the common standard assumptions, the desired indoor temperature based on the adaptive approach to thermal comfort, which is more properly climate-connected towards user expectations (Ferrari and Zanotto 2012a). These additional aspects are described in detail in the following sections.

6.1.1 Passive Cooling Strategies

Apart from the most advanced applications, the conventional simplest passive cooling strategies involve shading the transparent elements and ventilating the rooms during the night-time (Santamouris 2007). Even if passive cooling management can be currently optimized to obtain the maximum effectiveness, in this work the effect of these strategies is simulated according to a conservative approach, coherent with the likely behaviour of the occupants, as described below.

6.1.1.1 Shading

The shading strategy is modeled considering the likely activation of the devices by the occupants: therefore the condition for it application in the simulation software is the level of direct solar radiation arriving on the glazed surfaces, which, in order to take into account both the issues of glare and overheating, is set to 100 W/m^2 (Beccali and Ferrari 2003).

The shading devices, applied to every building exposure except for the Northern one, are external venetian blinds. Within the simulation software, the shading type needs to be represented with a decrease factor of the solar heat gains through the windows and with its position (internal or external), which influences the related thermal performance. For the venetian blinds is here assumed a shading factor of 30 %, which corresponds to a decrease in the solar heat gains of 70 %.

6.1.1.2 Night Ventilation

Night ventilation is a very effective cooling strategy, since it allows a significant decrease of the air temperature just before the beginning of the building occupancy time.

According to what stated in the norms EN ISO 13790 (2008) and UNI TS 11300-1 (2008), night ventilation is modelled between 23:00 and 07:00 during the cooling season (and only when the indoor operative temperature exceeds the cooling set-point), with an increase in the air-change rate of 5 h^{-1}, which is suggested as a value that is low but consistent with ventilation rates naturally achievable through single sided openings (Kolokotroni and Aronis 1999).

6.1.2 Indoor Set-Point Temperature

In common practice active systems are regulated according to air temperature sensors and the building simulation praxis refers to the same approach. However, since the comfort requirements from international standards, such as EN 15251 (2007) are expressed in terms of operative temperature, the case-study building set-point regulation is performed according to this value: thanks to this assumption, it is also possible to properly take into account the performances of the different building envelopes in contributing to overall thermal comfort sensation, i.e. in providing suitable surface radiant temperatures. These last ones, in case of unfavourable resulting values, are often responsible of an overusing the active climatization systems for correcting the indoor condition, air side, implying additional energy consumptions: this aspect is commonly neglected in building energy assessment, since the considered set-point simply refers to the air temperature.

The operative temperature, in fact, can be defined as a linear combination of air temperature and mean radiant temperature and can be calculated according to the following equation (EN ISO 7726 1998):

$$T_{op} = \frac{h_c T_{air} + h_r T_{mr}}{h_c + h_r} \tag{6.1}$$

where

T_{op} is the operative temperature [°C]
h_c is the heat transfer coefficient by convection [W/(m^2 K)]
T_{air} is the air temperature [°C]
h_r is the heat transfer coefficient by radiation [W/(m^2 K)]
T_{mr} is the mean radiant temperature [°C]

Since under usual indoor environmental conditions the two components, convective and radiant, can be considered equally effective on the operative temperature value, in the software TRNSYS T_{op} is calculated as the average between air and mean radiant temperature by default.

Therefore, T_{op} values of 20 °C for heating and 26 °C for cooling, according to what stated within EN 15251 (2007) for normal level of comfort expectations, are set for the energy need analysis of the case study.

Additionally, alternative T_{op} values are also considered for comparing the energy need analysis according to the adaptive approach, as an assessment method for more properly evaluating, with particular reference to the summer season, the building performances (Ferrari and Zanotto 2012b).

In this regard, since the CEN adaptive method provided in EN 15251 (2007) is valid for outdoor reference temperatures up to 30 °C, only its running mean temperature equation is considered for this study and assumed in the Dutch ATG standard (Van der Linden et al. 2006), which applies up to 33.5 °C therefore it is more proper for the Mediterranean climatic context. Additionally, the ATG formulation is widely adoptable, also for winter evaluations and for every kind of building (whether it is conditioned or not), while the CEN one is exclusively for summer season analysis in buildings without air-conditioning systems.

Hence, the ATG equations, in particular the ones suggested in order to get the 80 % acceptability that correspond to “good indoor conditions”, have been adopted.

The resulting adaptive temperature comfort ranges calculated for the considered locations are shown in Figs. 6.1, 6.2, 6.3, 6.4, 6.5, 6.6 and 6.7 compared to the

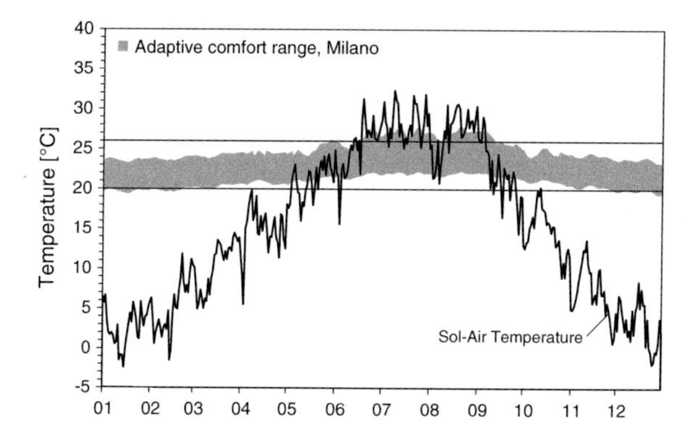

Fig. 6.1 Considered comfort ranges for the city of Milano

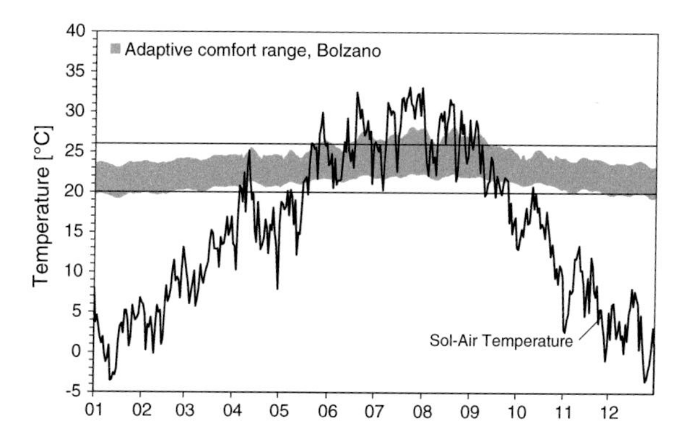

Fig. 6.2 Considered comfort ranges for the city of Bolzano

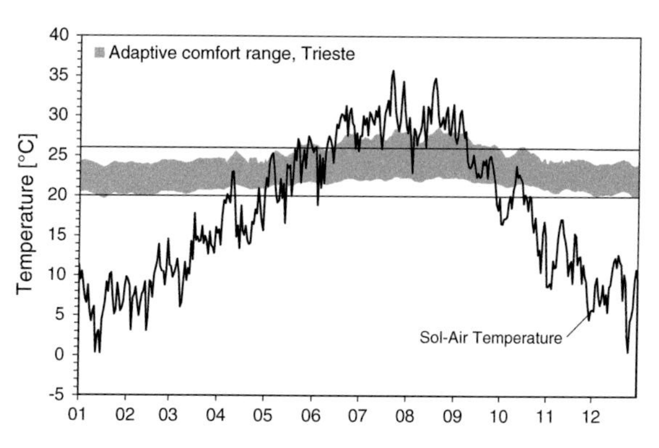

Fig. 6.3 Considered comfort ranges for the city of Trieste

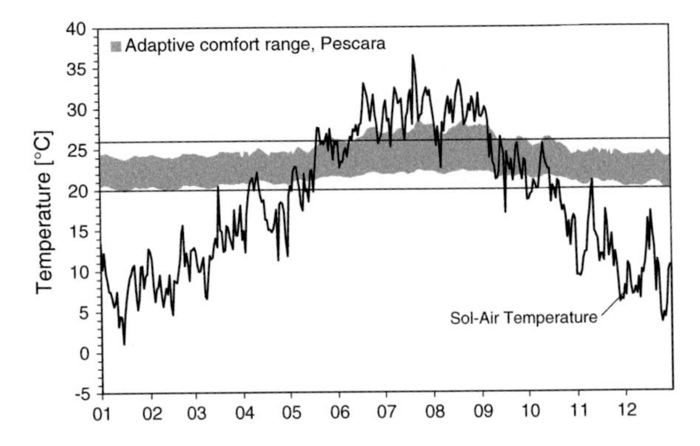

Fig. 6.4 Considered comfort ranges for the city of Pescara

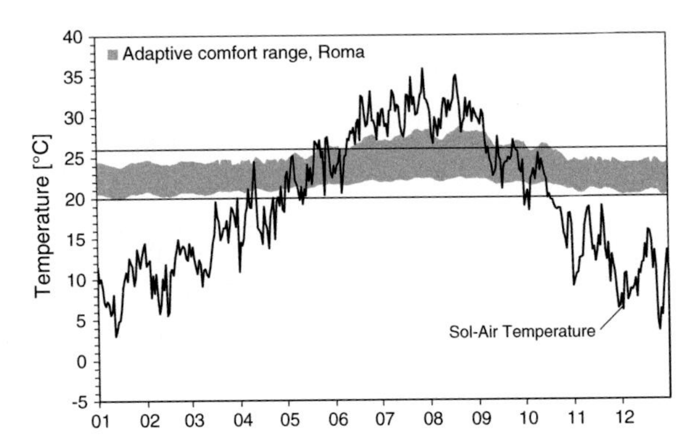

Fig. 6.5 Considered comfort ranges for the city of Roma

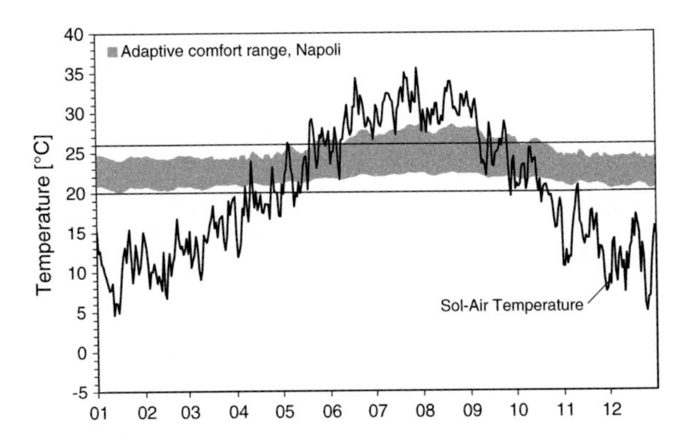

Fig. 6.6 Considered comfort ranges for the city of Napoli

conventional set-points. Additionally, to contextualize the climatic condition affecting building behaviour, also the related sol-air temperature (ASHRAE 2001) calculated on a horizontal surface is reported.

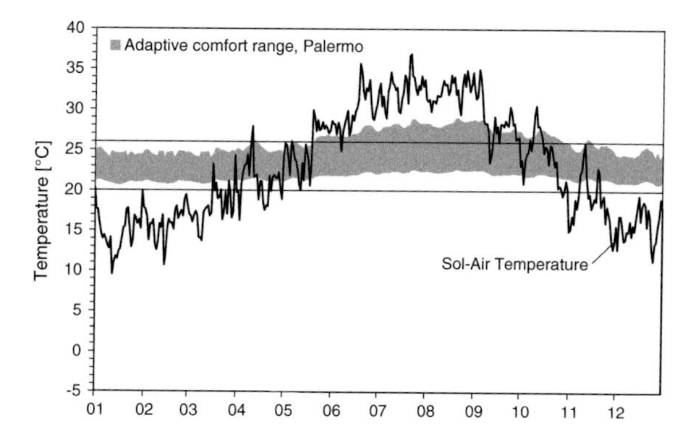

Fig. 6.7 Considered comfort ranges for the city of Palermo

6.2 Comparison of Buildings Performances

Here the hourly dynamic simulation results are analysed in detail, in order to compare the behaviour of the different building models in terms of both free-floating indoor resulting temperature (without active climatization devices) and heating or cooling sensible thermal need (by supposing an active system able to satisfy the temperature set-points requirement, based on the occupancy time-schedule that in this case refers to the office working days).

In order to reduce the amount of data to display, representative cases taken from the set of simulations are analysed, with reference to a typical winter week (January 8th–14th) and a typical summer week (July 9th–15th), considering a single building exposure (South) placed in an average climatic location among the selected ones (Rome).

6.2.1 *Winter Week*

Concerning the winter free-floating behaviour, the operative temperature values in Fig. 6.8 show that the new buildings, characterized by lower average U-value, have obviously the advantageous higher minimum values, but it is interesting to note that, independently on the age, the buildings with heavier constructions and smaller windows (new conventional, 60/80 conventional and traditional) are characterized by smaller daily variability than the lighter and more glazed buildings. Moreover, these last (new glazed and 60/80 sandwich) reveal higher temperature peaks, also bringing to overheating.

Figure 6.9 reports the hourly heating need values for the considered constructions during the same week according to the conventional set-point temperature (the winter adaptive set-point option does not imply significant differences because, for the selected climatic context, 20 °C is almost compatible with the adaptive temperature range). The trends of the thermal load bring to the remarks consistently

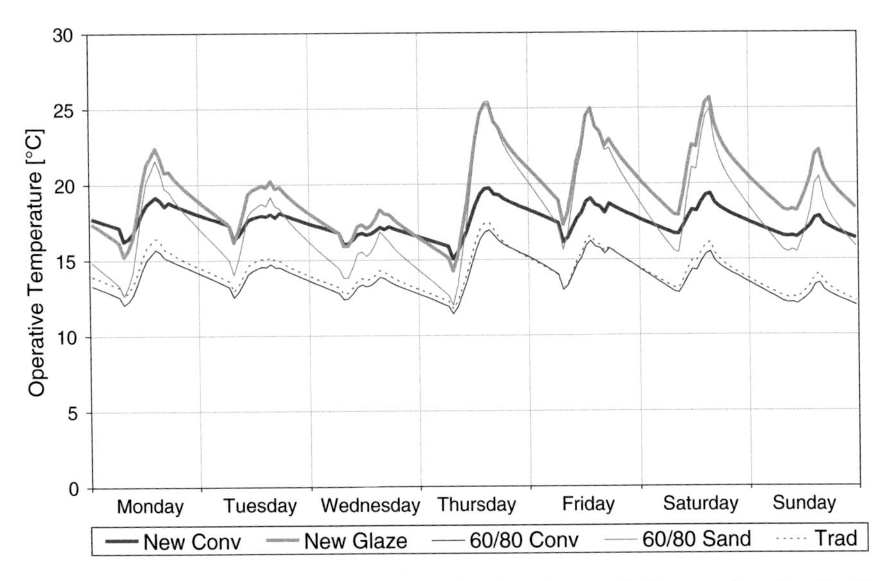

Fig. 6.8 Free-floating operative temperatures during the winter week (Rome, orientation South)

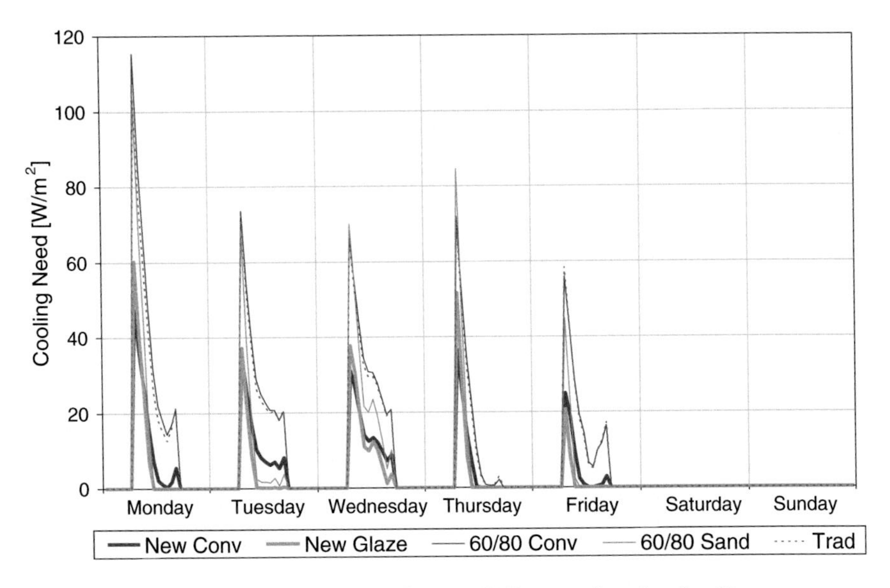

Fig. 6.9 Hourly heating need during the winter week (Rome, orientation South)

with the free-floating analysis: on one hand older buildings, with higher U-values, have higher maxima in the mornings because during the night-time the indoor operative temperature decreases a lot, while the lighter and more glazed constructions are related to lower needs (even close or equal to zero) during the central hours of the day, due to the effect of direct solar radiation.

6.2.2 Summer Week, Basis

The summer free-floating temperature trends are shown in Fig. 6.10. From the graph it can be seen that once again the lighter buildings are the ones characterized by a stronger daily temperature oscillation, also due to the larger glazed surface (which is not shaded in this case). Additionally, it is interesting to note that the older conventional constructions perform the lower indoor temperatures, since with lower envelope U-values it is easier to lose the indoor heat.

The cooling need trends are reported in Figs. 6.11 and 6.12, one based on the conventional set-point and the other based on the adaptive one: by comparing them, the different magnitude of cooling need assessment depending on the considered approach can be strongly appreciated, since the desired adaptive temperature in the selected week is always significantly higher than 26 °C.

Both cases show the higher maxima, in particular of the lighter constructions, during the central hours of the day because of the direct solar radiation through the unshaded large windows. The new conventional building, among the heavier ones, has higher loads at the beginning of the occupied hours, since it cannot loose during the night-time all the heat stored during the day. This last remark is highlighted at the beginning of the week, because of the rise in the indoor temperature during the unoccupied (i.e. unconditioned) weekends.

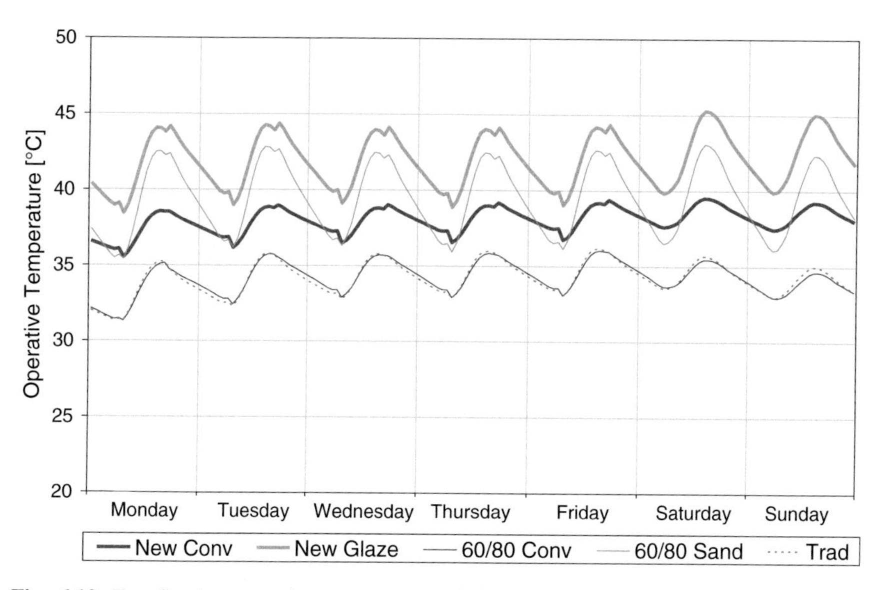

Fig. 6.10 Free-floating operative temperatures during the summer week (Rome, orientation South)

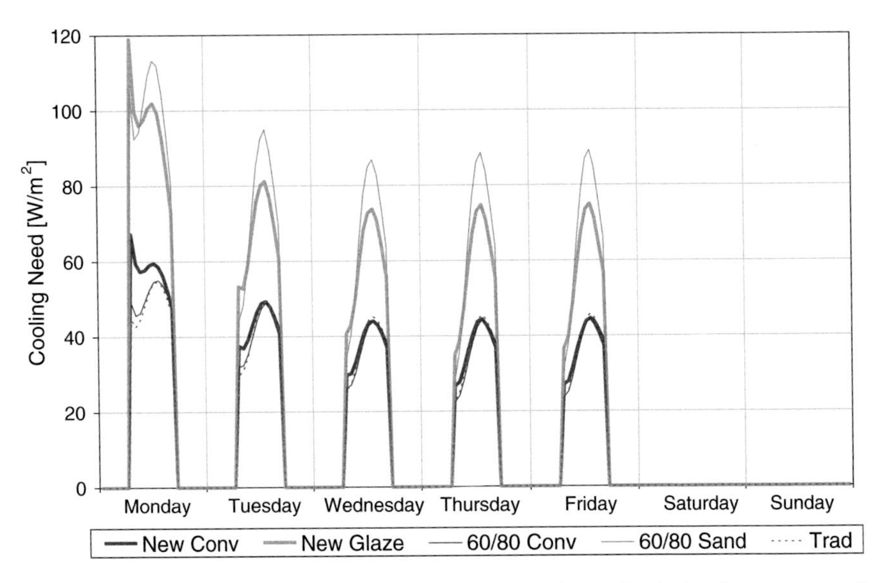

Fig. 6.11 Hourly cooling need according to the conventional set-point during the summer week (Rome, orientation South)

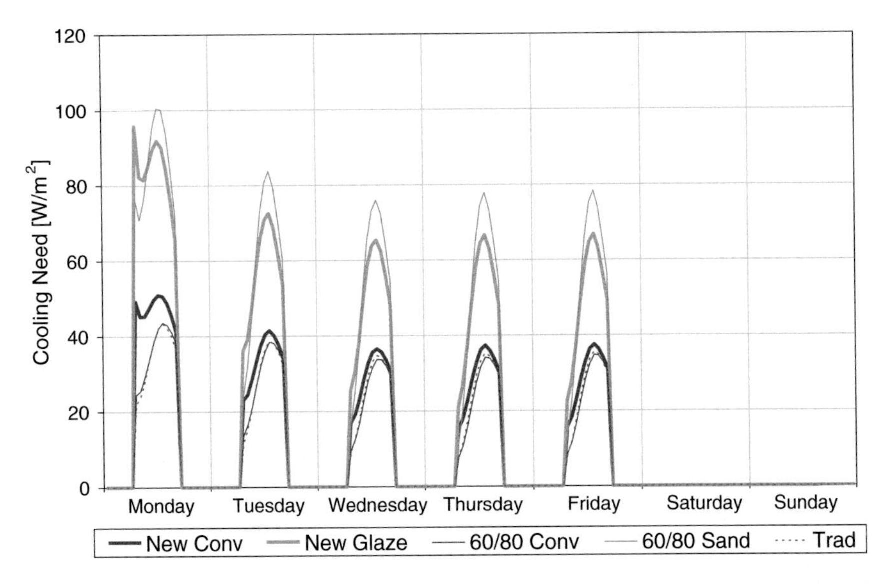

Fig. 6.12 Hourly cooling need according to the adaptive set-point during the summer week (Rome, orientation South)

6.2.3 Summer Week, with External Shading

Figure 6.13 shows the summer operative temperature trends in free floating, and Figs. 6.14 and 6.15 the cooling load related to the two set-point options, when external venetian blinds are used as shading device. In all cases the maximum values are obviously less pronounced than in the previous evaluations, but in particular with reference to the lighter constructions.

6.2.4 Summer Week, with Night Ventilation

Figure 6.16 shows the summer operative temperature trends in free floating when night ventilation is adopted as a free cooling strategy. The daily average temperature decreases and the daily variability increases in all cases, thanks to the cooler conditions during the night-time, but the most important remark regards the fact that the behaviour of the new conventional building tends to become more similar to the one of the older heavyweight constructions. This is due to the fact that night ventilation allows the internal mass to lose the heat during the unoccupied hours. The same can be said, even if with lower extent, of the new glazed construction, which tends to perform like the older largely glazed one (60/80 sandwich).

The cooling need trends in Figs. 6.17 and 6.18 show that the load in the first hours of the occupied time is much smaller respect the previous cases thanks to the

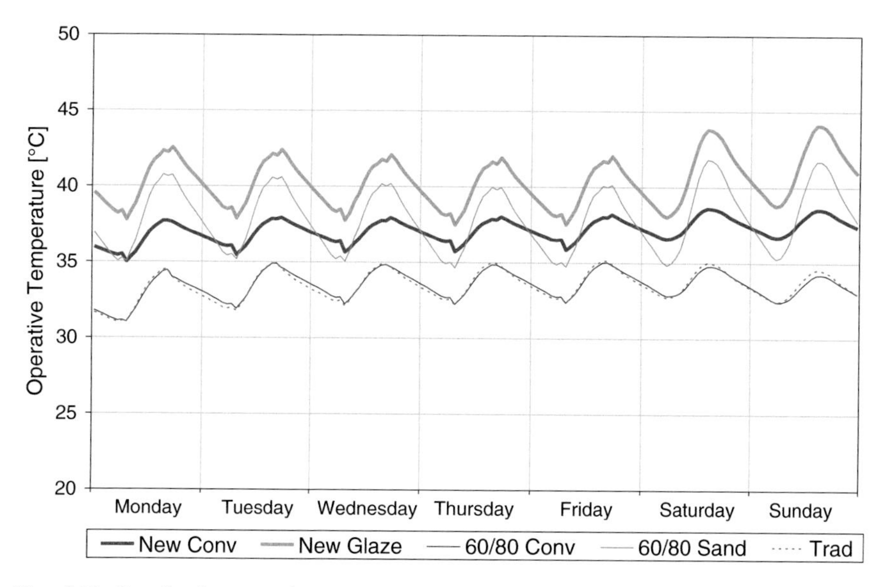

Fig. 6.13 Free-floating operative temperatures during the summer week, in case of external shading (Rome, orientation South)

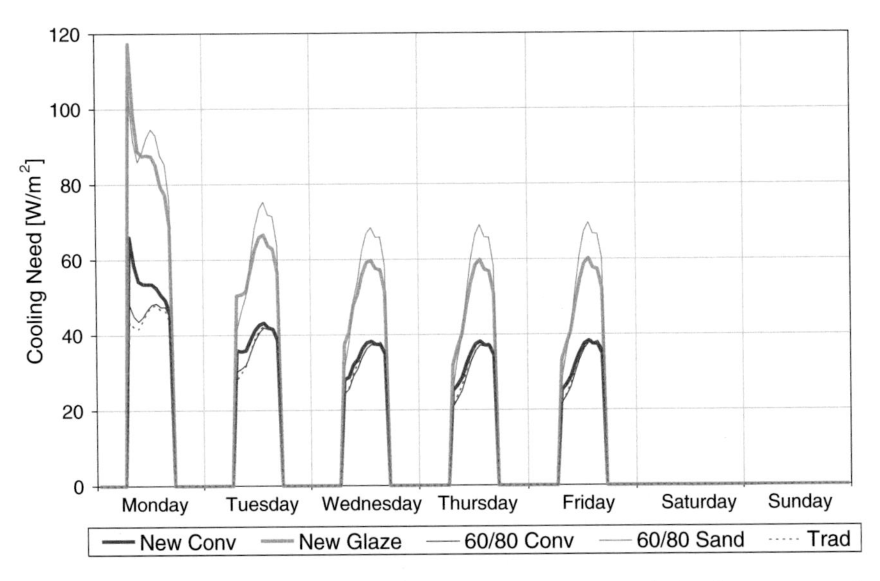

Fig. 6.14 Hourly cooling need according to the conventional set-point during the summer week, with shading (Rome, orientation South)

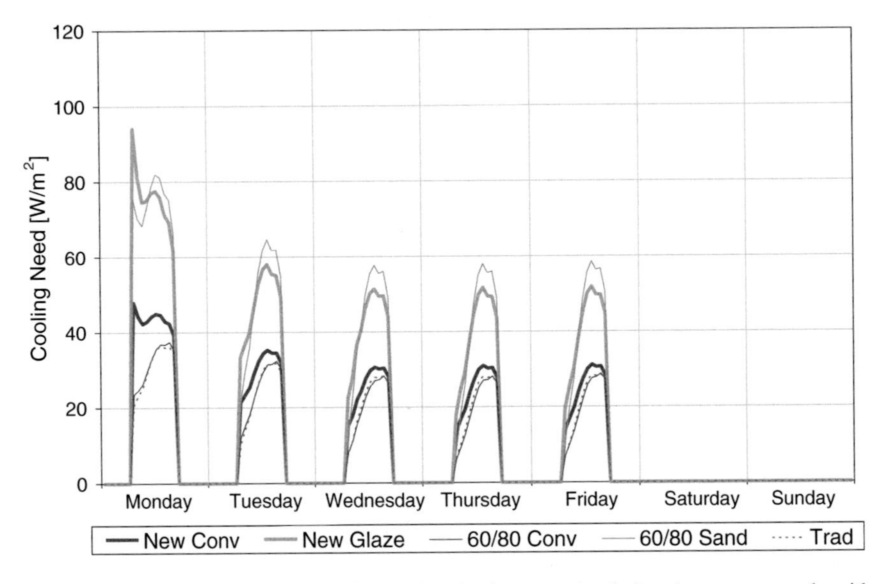

Fig. 6.15 Hourly cooling need according to the adaptive set-point during the summer week, with shading (Rome, orientation South)

night-time temperature decrease, in particular regarding the first day of the week, and therefore the differences between the new conventional building and the other heavy constructions almost disappear.

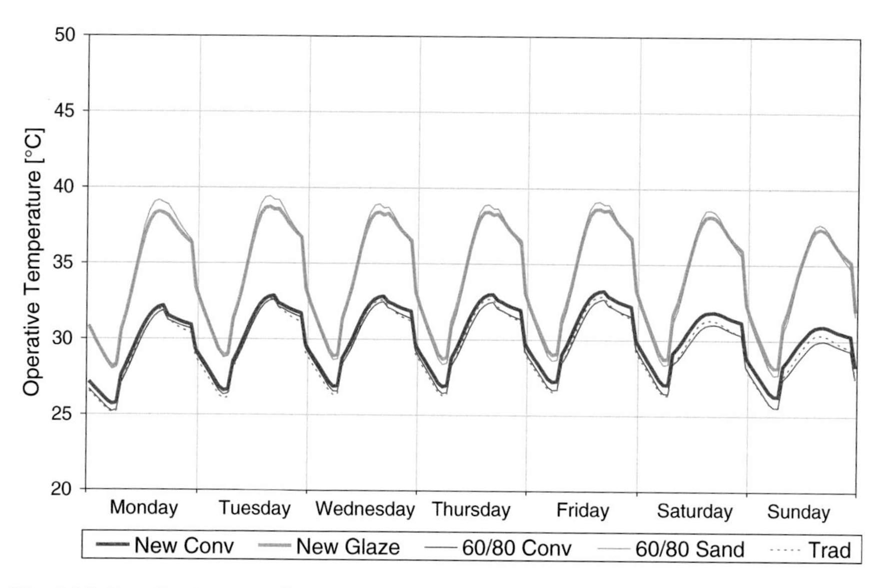

Fig. 6.16 Free-floating operative temperatures during the summer week, with night ventilation (Rome, orientation South)

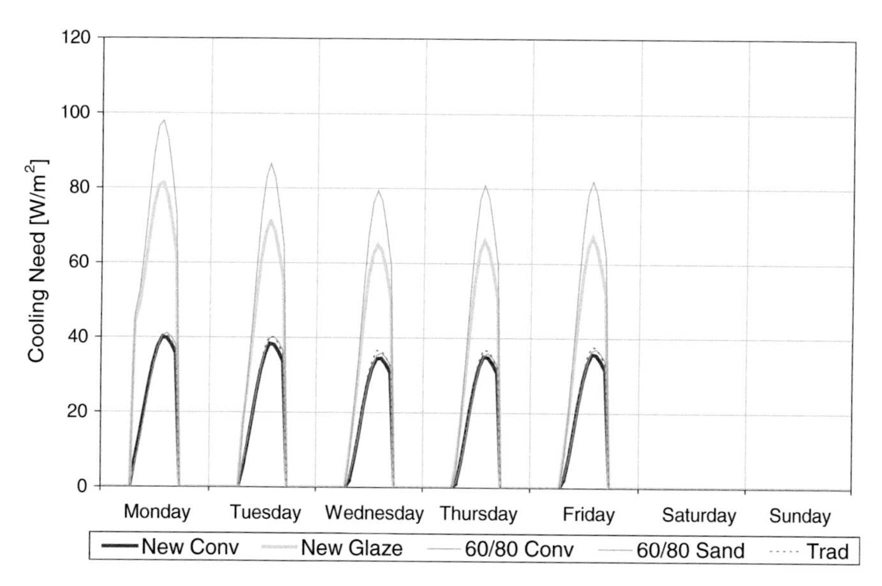

Fig. 6.17 Hourly cooling need according to the conventional set-point during the summer week, with night ventilation (Rome, orientation South)

When the adaptive set-point is used, the decrease in the cooling need due to the additional ventilation is more pronounced, in particular for the heavier construction: this is once again due to the strong effect of heat loss from the massive structures caused by the additional discharge rate during the night-time.

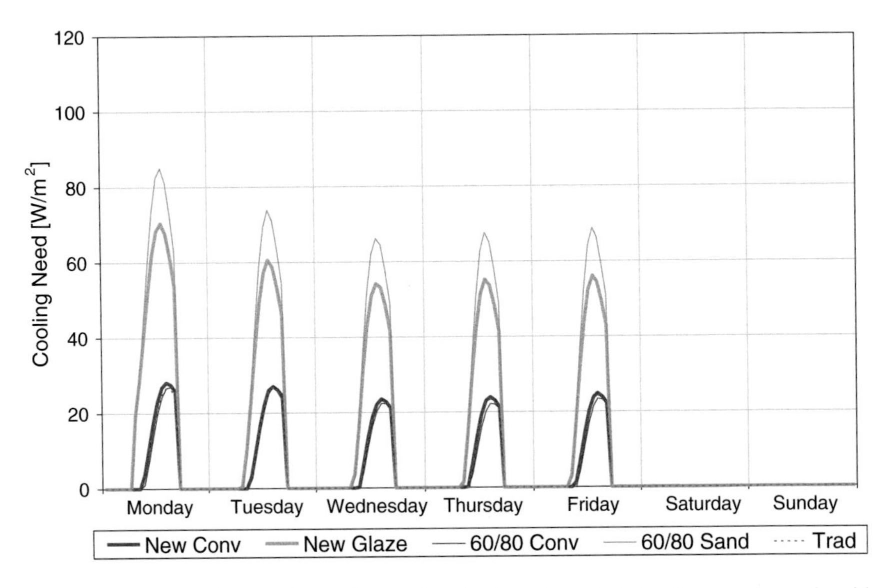

Fig. 6.18 Hourly cooling need according to the adaptive set-point during the summer week, with night ventilation (Rome, orientation South)

6.2.5 Summer Week, with External Shading and Night Ventilation

Figure 6.19 reports the free-floating operative temperature calculated for the summer week when both the shading and the night ventilation strategies are considered: the effects already seen for the single passive cooling applications are combined, as well under the point of view of the cooling need (Figs. 6.20 and 6.21).

6.3 Effect of a Climate-Connected Set-Point to the Seasonal Cooling Needs

Since the adaptive indices have been developed according to the actual building users' thermal sensations and preferences (Ferrari and Zanotto 2012a), in this work the adaptive comfort temperature represents the climatization system set-point as autonomously managed by the occupants, according to the external climate. This aspect, as previously saw, can be translated into a cooling need assessment quite different respect the one referred to the conventional constant temperature setting. Moreover, different building solutions can perform differently in this field.

In Figs. 6.22, 6.23, 6.24, 6.25, 6.26, 6.27 and 6.28, the seasonal cooling reductions connected to the introduction of the variable adaptive set-point in summer are shown for all the cases of the matrix.

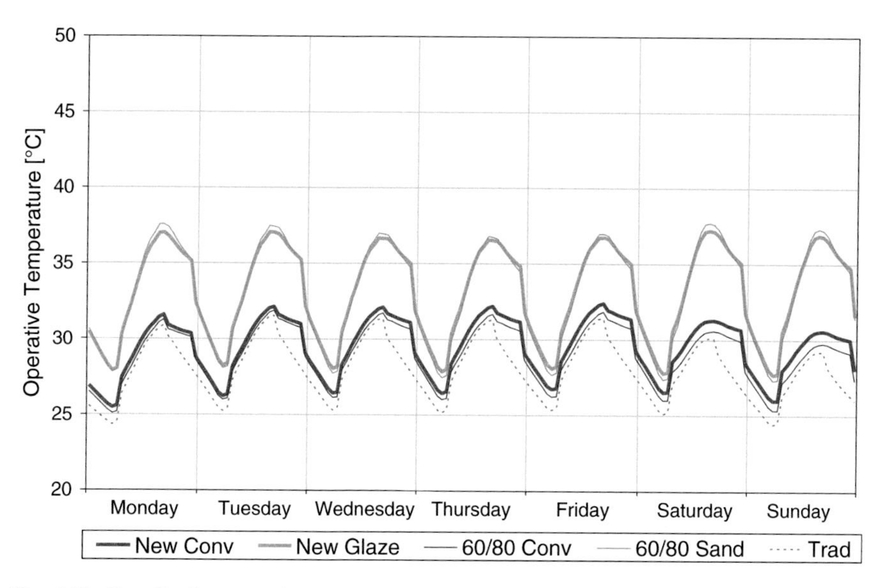

Fig. 6.19 Free-floating operative temperatures during the summer week, with shading and night ventilation (Rome, orientation South)

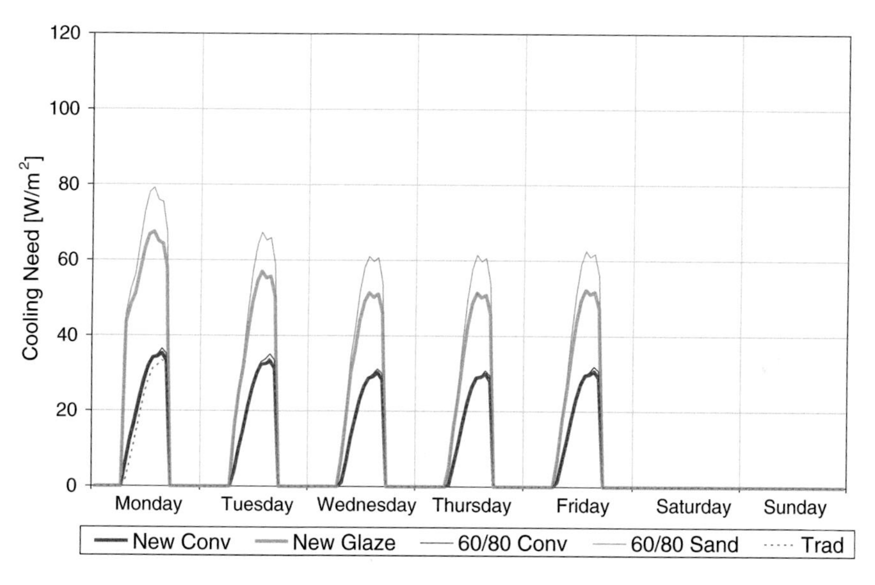

Fig. 6.20 Hourly cooling need according to the conventional set-point during the summer week, with shading and night ventilation (Rome, orientation South)

During the cooling season, the cases that reveal greatest differences based on the adaptive temperature set-point are the massive buildings, in particular the older ones, which are not provided with any insulation layer. This can be clearly seen in

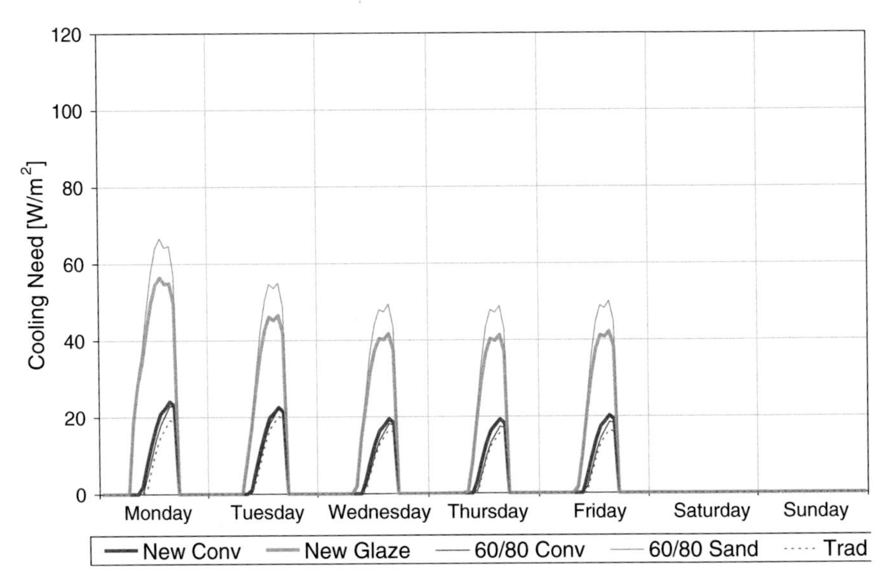

Fig. 6.21 Hourly cooling need according to the adaptive set-point during the summer week, with shading and night ventilation (Rome, orientation South)

the base case and in the shaded case, while only the night ventilation strategy, allowing the loss of the stored heat, strongly reduces the calculated need of the new conventional building.

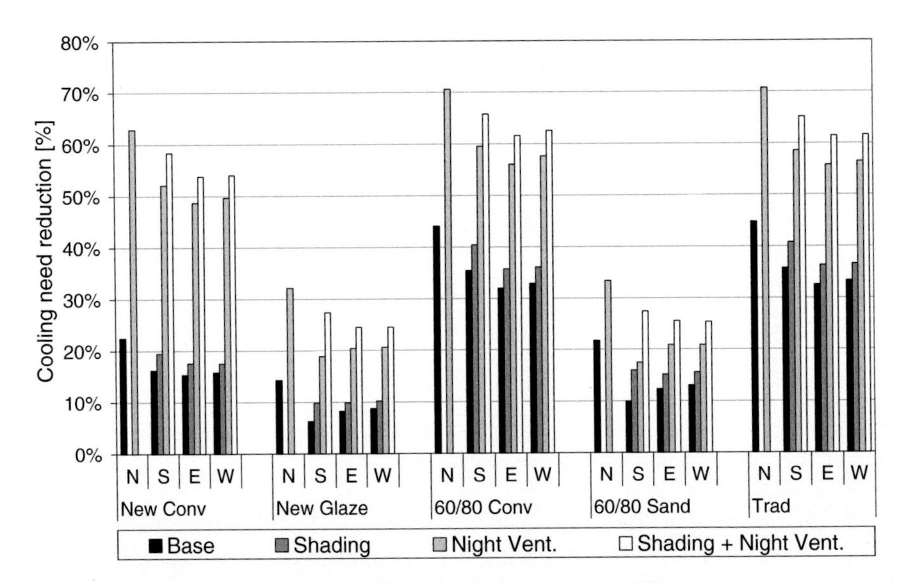

Fig. 6.22 Cooling need reduction based on the adaptive set-point. Milano

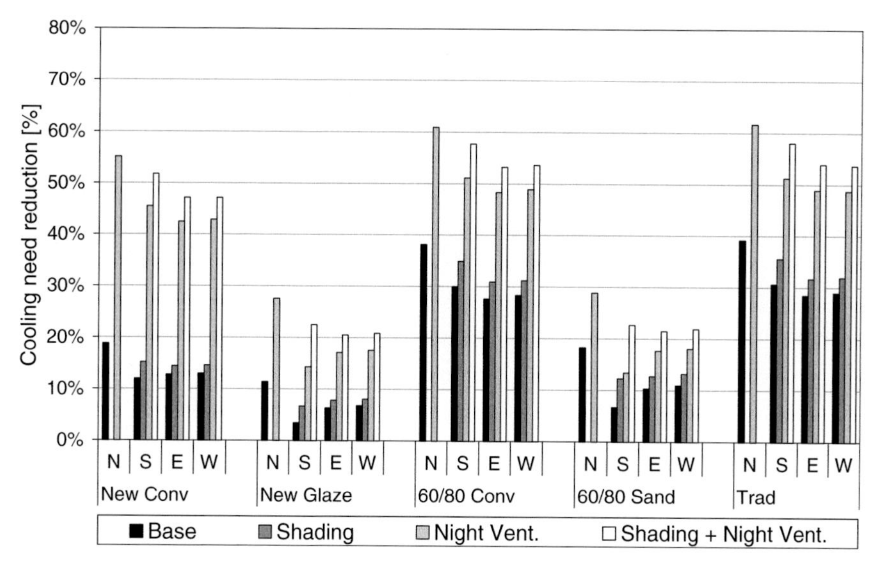

Fig. 6.23 Cooling need reduction based on the adaptive set-point. Bolzano

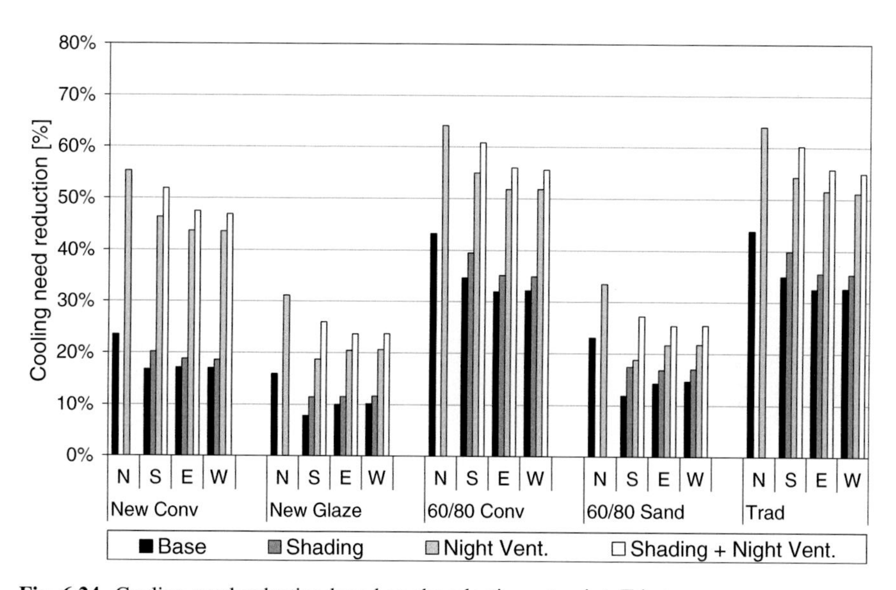

Fig. 6.24 Cooling need reduction based on the adaptive set-point. Trieste

It is also interesting to see, by comparing all the graphs, that the reduction in the cooling need assessed for the heavier constructions tend to decrease as the climate becomes warmer, while in case of the lighter buildings the trend is inverse. This is

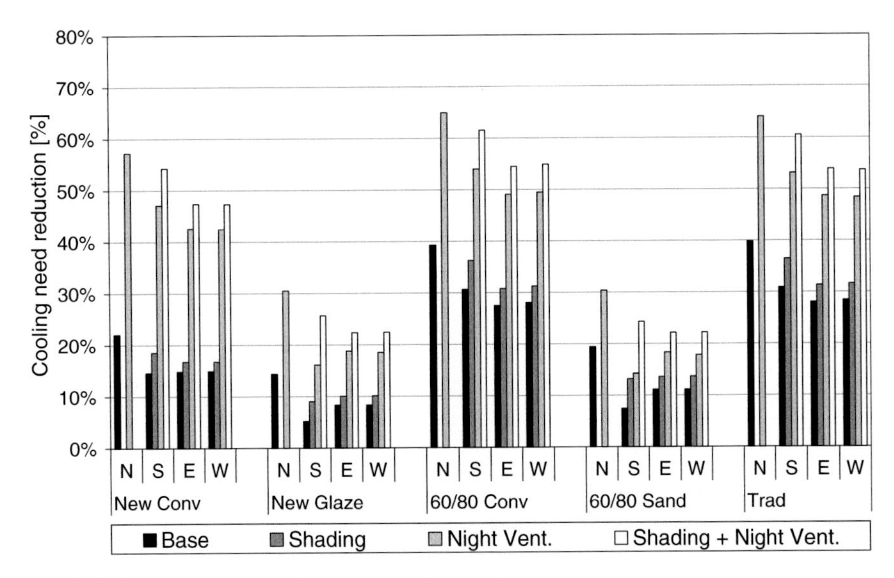

Fig. 6.25 Cooling need reduction based on the adaptive set-point. Pescara

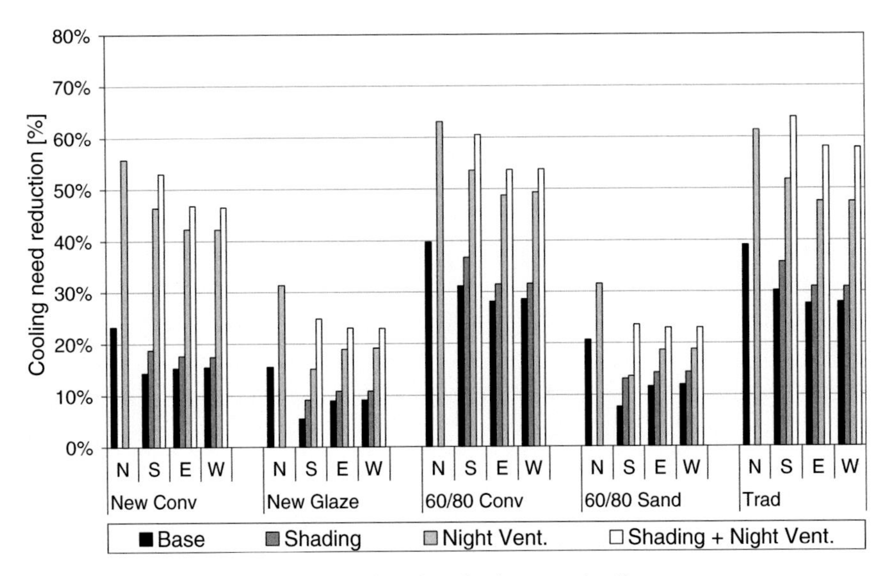

Fig. 6.26 Cooling need reduction based on the adaptive set-point. Roma

due to the fact that in warmer climates there is larger gap between the conventional set-point temperature and the user expected one: this aspect also highlights the peculiarity of the southern European context.

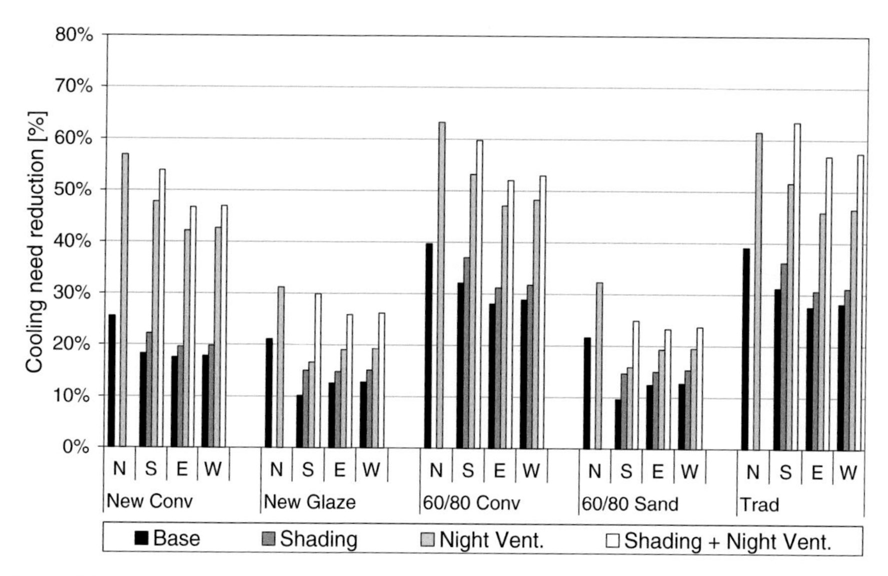

Fig. 6.27 Cooling need reduction based on the adaptive set-point. Napoli

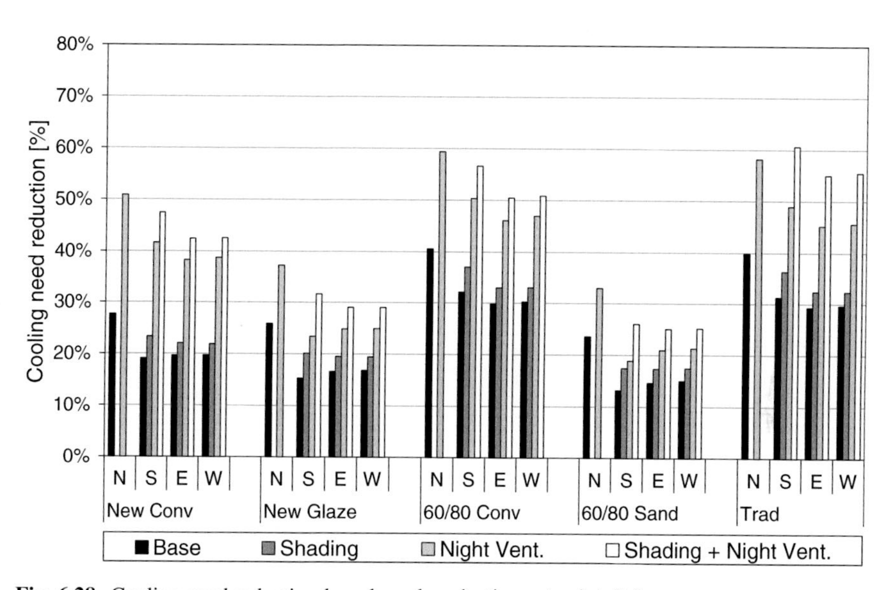

Fig. 6.28 Cooling need reduction based on the adaptive set-point. Palermo

References

ASHRAE, *Handbook of Fundamentals* (American Society of Heating, Refrigerating and Air-conditioning Engineers, Atlanta, 2001)

M. Beccali, S. Ferrari, Energy saving and comfort in office buildings: performances of different combinations of window technologies and lighting control strategies, in Proceedings of the 20th Conference on Passive and Low Energy Architecture, Santiago, Chile, 9–12 Nov 2003

EN 15251, *Indoor Environmental Input Parameters for Design and Assessment of Energy Performance of Buildings Addressing Indoor Air Quality, Thermal Environment, Lighting and Acoustics* (European Committee of Standardization, Brussels, 2007)

EN ISO 7726, *Ergonomics of the Thermal Environment—Instruments for Measuring Physical Quantities* (European Committee of Standardization, Brussels, 1998)

EN ISO 13790, *Energy Performance of Buildings—Calculation of Energy Use for Space Heating and Cooling* (European Committee of Standardization, Brussels, 2008)

S. Ferrari, V. Zanotto, Adaptive comfort: analysis and application of the main indices. Build. Environ. **49**, 25–32 (2012a)

S. Ferrari, V. Zanotto, Office buildings cooling need in the Italian climatic context: assessing the performances of typical envelopes. Energy Procedia **30**, 1099–1109 (2012b)

S.A. Klein, W.A. Beckman, J.W. Mitchell et al., *TRNSYS—A Transient System Simulation Program User Manual* (The solar energy Laboratory—University of Wisconsin, Madison, 2007)

M. Kolokotroni, A. Aronis, Cooling energy reduction in air-conditioned offices by using night ventilation. Appl. Energ. **63**, 241–253 (1999)

M. Santamouris, *Advances in Passive Cooling* (Earthscan, London, 2007)

UNI TS 11300-1, *Determinazione del fabbisogno di energia termica dell'edificio per la climatizzazione estiva e invernale* (Ente Nazionale Italiano di Unificazione, Milano, 2008)

A.C. Van der Linden, A.C. Boerstra, A.K. Raue AK et al., Adaptive temperature limits: a new guideline in the Netherlands. A new approach for the assessment of building performance of with respect to thermal indoor climate. Energ. Build. **38**, 8–17 (2006)

Chapter 7
Climate-Related Assessment of Building Energy Needs

Abstract This chapter reports the analysis of the correlation between the seasonal thermal energy needs of different building solutions and their climatic locations, assuming the variety of the Italian context taken as representative of southern Europe. The case-studies, simulated through a detailed dynamic tool, are representative of common practices from three main construction ages (newly built, 1960/80 and very old), and therefore include both different envelope characteristics and window percentages. Consideration is also given to the implementation of passive cooling strategies, such as shading and night ventilation, and to the adoption of variable set-point temperatures, directed towards climate-related user expectations, i.e. adaptive comfort. The graphic representation, according to the heating and cooling degree days adopted for supporting the climate-related analysis, can also be useful for extending the assessment to other locations within the same climatic area through simple interpolation. The data sheets reporting the detailed results of the case-studies, on the other hand, can be used directly for assessing the building energy needs of similar locations.

Keywords Climate-related building energy need · Southern european building context · Specific building energy need values · Indoor operative temperature requirement · Passive cooling strategies · Adaptive set-point

7.1 Assessing Building Energy Needs

Research studies conducted by the authors (Ferrari and Zanotto 2011, 2012b) aimed at assessing the thermal behaviour of typical buildings within the wide range of the Italian climatic conditions, which can be representative of southern Europe. Starting from a simple reference building shape the case-studies were defined as representative of likely practices from three main construction ages: newly built, 1960/80 and very old.

For every construction age, "conventional" vertical envelope solutions have been taken into account, with masonry external walls and standard windows size.

S. Ferrari and V. Zanotto, *Building Energy Performance Assessment in Southern Europe*, PoliMI SpringerBriefs,
DOI 10.1007/978-3-319-24136-4_7

Moreover, alternative more glazed and lighter solutions have been considered for the contemporary and 1960/80 ages (detailed description in Chap. 5).

The set of the building models, simulated by the means of the detailed dynamic software TRNSYS 16 (Klein et al. 2007), also included the cases of passive cooling strategies implementation, such as shading and night ventilation, and of the adoption of variable set-point temperatures based on the adaptive approach to thermal comfort, more properly climate-connected towards user expectations (Ferrari and Zanotto 2012a) (detailed description in Chap. 6).

In Tables 7.1 and 7.2 the main building envelope elements characterising the case studies (i.e. external walls and glazed surfaces) are reported with reference to the chosen seven locations (Bolzano, Milano, Trieste, Pescara, Roma, Napoli and Palermo).

The rooms and their openings are located on the two main façades of the building in order to appreciate, through only two rotating options of the models, the performances of the four exposures covering the main orientation cases (North, South, East and West). Hence, simulation outputs of energy need necessary to reach the required operative temperature, were obtained for the building thermal zones having the same exposure, allowing a performance comparison also based on the different exposures.

Additionally, an analysis of the correlation between seasonal thermal energy needs of different building solutions and their climatic locations has been provided and reported in this chapter. The climatic-related analysis, beside the performance comparisons reported in the previous Chap. 6, could be taken into account for evaluating the implications when selecting the solutions in the design phase.

Moreover, the specific results (i.e. per square meter of floor space facing each exposure) could be extrapolated for estimating, in first approach, the thermal energy need of buildings having similar solutions even with different shapes. These estimations can be extended also to other locations within the same climatic area, through graphical interpolation, once the heating and cooling degree days of the considered location have been calculated consistently with the procedure assumed in this study.

7.2 Climate-Related Analysis

Dynamic simulation software always requires hourly data. Within TRNSYS (Klein et al. 2007) a database of climatic files (Test Reference Year data) is provided and the ones referred to the locations selected for this study have been used for the simulations. According to the same TRY data, heating and cooling degree-day values (HDD and CDD respectively) have been calculated and adopted in the present study.

$$HDD = \sum T_{base} - T_{ext,day}, \quad \text{when} \quad T_{ext,day} < T_{base} \tag{7.1}$$

Table 7.1 Characteristics of the external walls for the different solutions

	Thermal insulation thickness [cm]	U-value [W/(m^2 K)]	Heat capacity [kJ/(m^2 K)]
New conventional	7.0 (MI/TS)–3.5 (PA)	0.33 (MI/TS)–0.47 (PA)	~260
New glazed	8.5 (MI/TS)–5.0 (PA)	0.33 (MI/TS)–0.47 (PA)	~251
1960/80 conventional	–	0.98	263
1960/80 sandwich largely glazed	10	0.36	53
Old traditional	–	1.08 (MI/PE)–0.98 (RM/PA)	818 (MI/PE)–587 (RM/PA)

Table 7.2 Characteristics of the glazed surfaces for the different solutions

	Window U-value [W/(m^2 K)]	Solar heat gain coefficient (g)
New conventional	1.77 (MI/TS)–2.81 (PA)	0.60 (MI/TS)–0.65 (PA)
New glazed	1.28 (MI/TS)–2.16 (PA)	0.39 (MI/TS)–0.41 (PA)
1960/80 conventional	2.91	0.76
1960/80 sandwich largely glazed	2.91	0.76
Old traditional	2.73	0.76

$$CDD = \sum T_{ext,day} - T_{base}, \quad \text{when} \quad T_{ext,day} < T_{base} \tag{7.2}$$

where

T_{base} is the base internal temperature [°C], assumed as 20 °C for the heating season (HDD) and 26 °C for the summer season (CDD)

T_{day} is the daily average external temperature [°C]

In particular, the external temperature ($T_{ext,day}$) here adopted for the calculation of the reference degree-days is the sol-air temperature, which is normally defined according to Eq. 1.16 (ASHRAE 2001). In this way, besides the air temperature, the direct solar radiation of the site affecting the building thermal needs can be taken into account in first analysis. For the present study, the simplified form of Eq. 7.3 was adopted, considering the solar radiation on a horizontal surface with average absorptivity.

$$T_{sa,e} = T_{air,e} + \frac{\alpha I}{h_e} \tag{7.3}$$

where

$T_{sa,e}$ is the sol-air temperature [°C]

$T_{air,e}$ is the outdoor air temperature [°C]

α is the absorption coefficient of the surface, assumed as 0.6

I is the global solar radiation on the surface [W/m^2]

h_e is the surface heat transfer coefficient [W/(m^2 K)], assumed as 25 W/(m^2 K)

Therefore, the calculated degree-day values are here referred to as HDD_{sa} in case of the heating season and as CDD_{sa} in case of the cooling season, as shown in Table 7.3 and Fig. 7.1.

Hence, climatic-related representations of the different cases have been provided and analysed.

As expected, the climatization need has a very strong correlation with the climatic conditions of the building location, which can be safely approximated to a linear one in most cases. The slope value of this correlation equation, however, strongly depends on the building solution.

The results regarding the heating season (Fig. 7.2) show that the correlation with the climatic parameter is higher in case of the older buildings, which are characterized by higher U-values.

Table 7.3 Heating and cooling degree-days calculated for the selected locations

	Milano	Bolzano	Trieste	Pescara	Roma	Napoli	Palermo
HDD_{sa}	2529	2472	1750	1595	1397	1234	551
CDD_{sa}	159	206	313	325	390	436	643

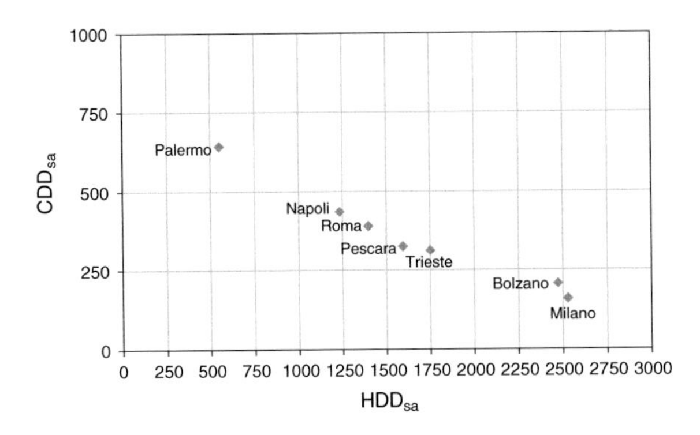

Fig. 7.1 Correlation between the heating and cooling degree-days calculated for the selected locations

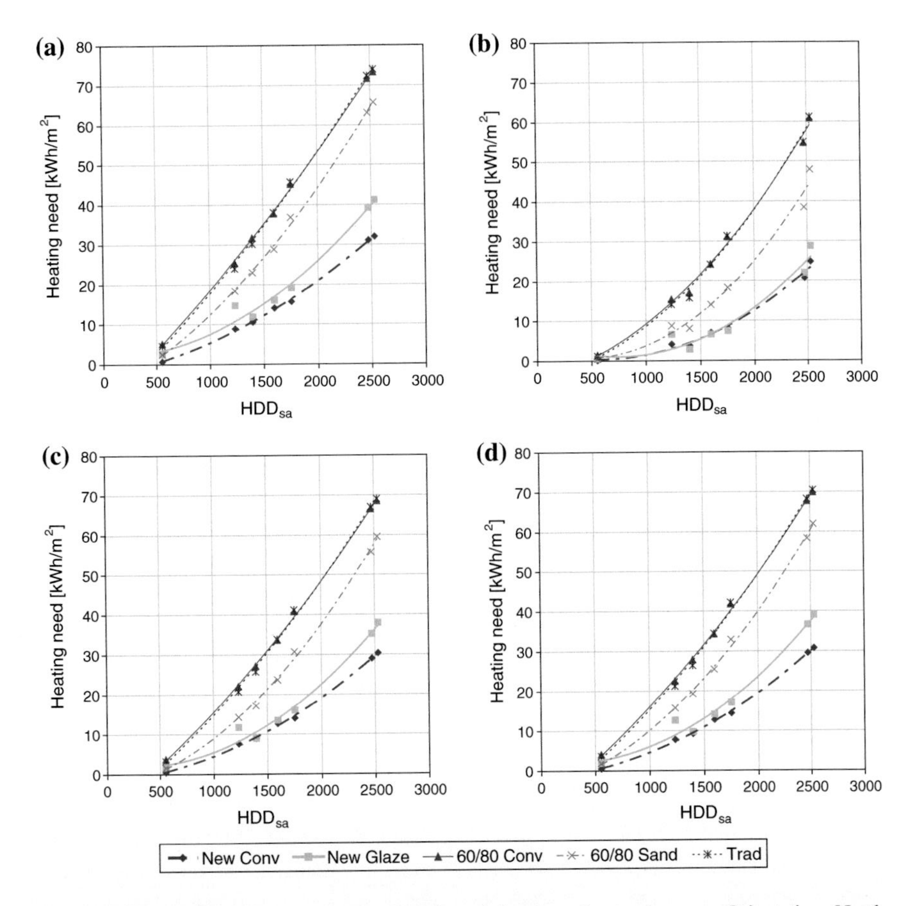

Fig. 7.2 Correlation between heating need and heating degree-days. **a** Orientation North. **b** Orientation South. **c** Orientation East. **d** Orientation West

The results regarding the cooling season (Fig. 7.3) show that the cooling need is always higher in case of the lighter buildings. Regarding the correlation equation, then, these buildings are characterized by a higher y-intercept and therefore by a higher cooling need "baseline". The 60/80 sandwich building (which is the one with the lowest thermal mass), moreover, shows very high slope values, that are often double than the ones of the other buildings.

Figures 7.4, 7.5 and 7.6 report the correlation between cooling degree-days and cooling need in case of application of the considered passive cooling strategies. It can be seen that as the passive cooling strategies are progressively introduced the correlation with the climatic parameter decreases in every building, as well as the y-intercept.

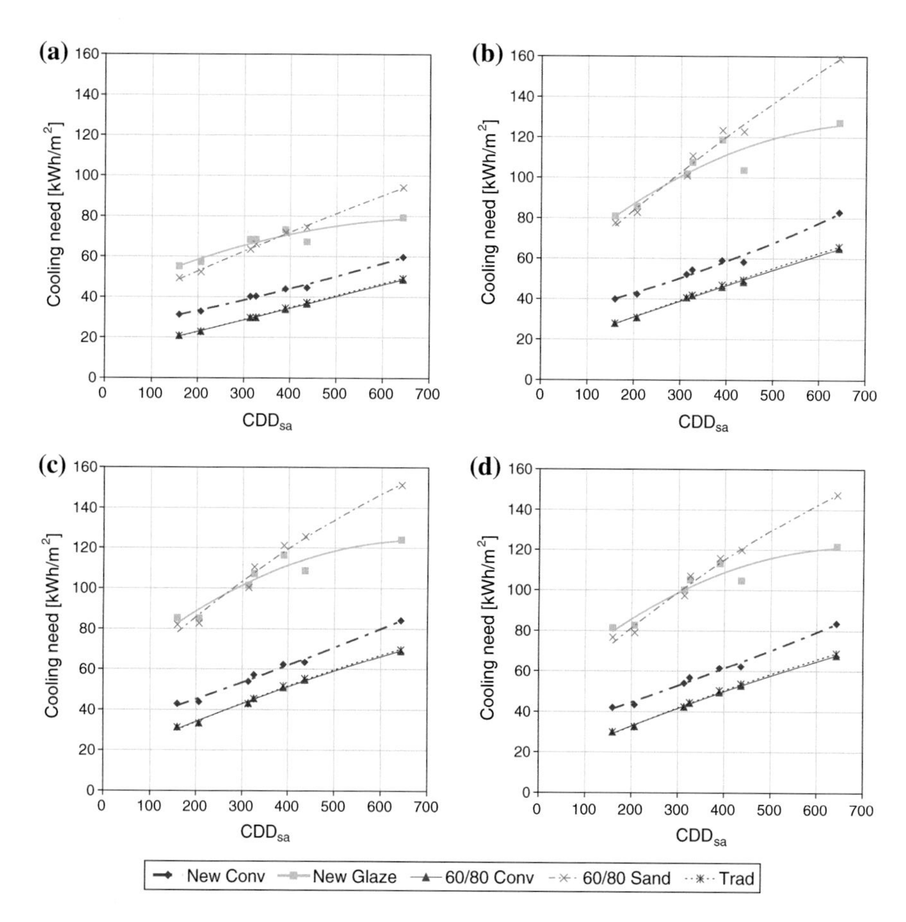

Fig. 7.3 Correlation between cooling need and cooling degree-days. **a** Orientation North. **b** Orientation South. **c** Orientation East. **d** Orientation West

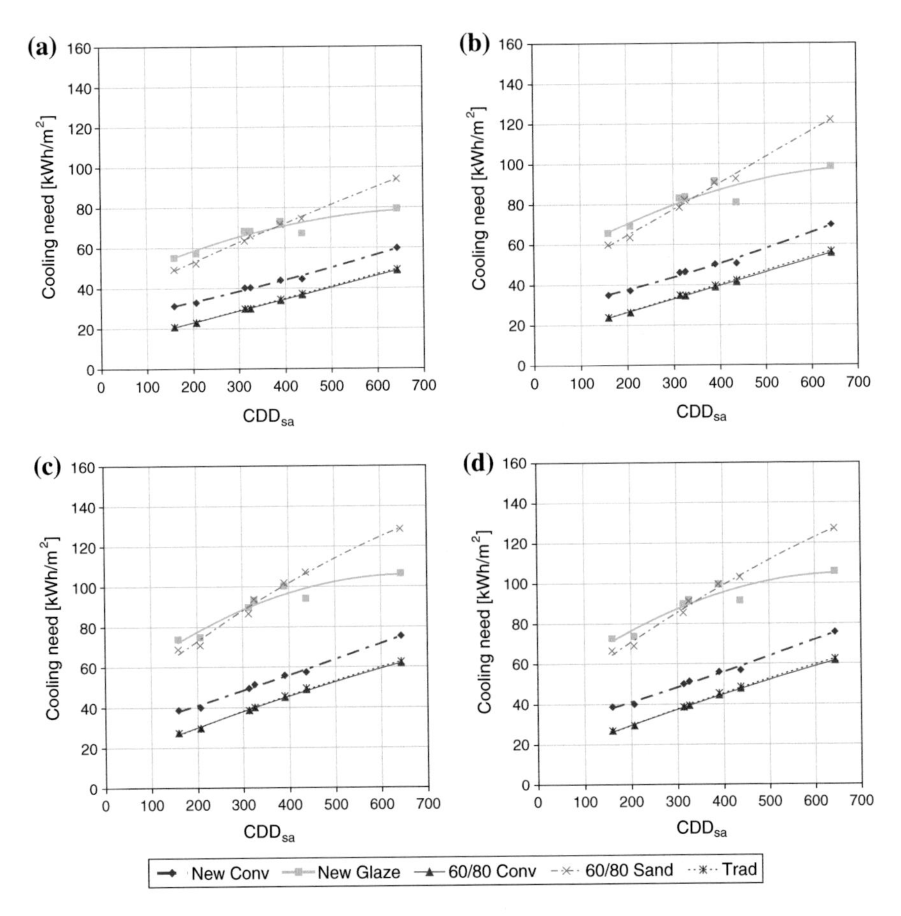

Fig. 7.4 Correlation between cooling need and cooling degree-days, with external shading. **a** Orientation North. **b** Orientation South. **c** Orientation East. **d** Orientation West

Finally, as it can be seen in Figs. 7.7 and 7.8, the previous trends increase if also the adaptive set-point is considered. In particular, passive cooling strategies show to be more effective when a variable temperature set-point is considered.

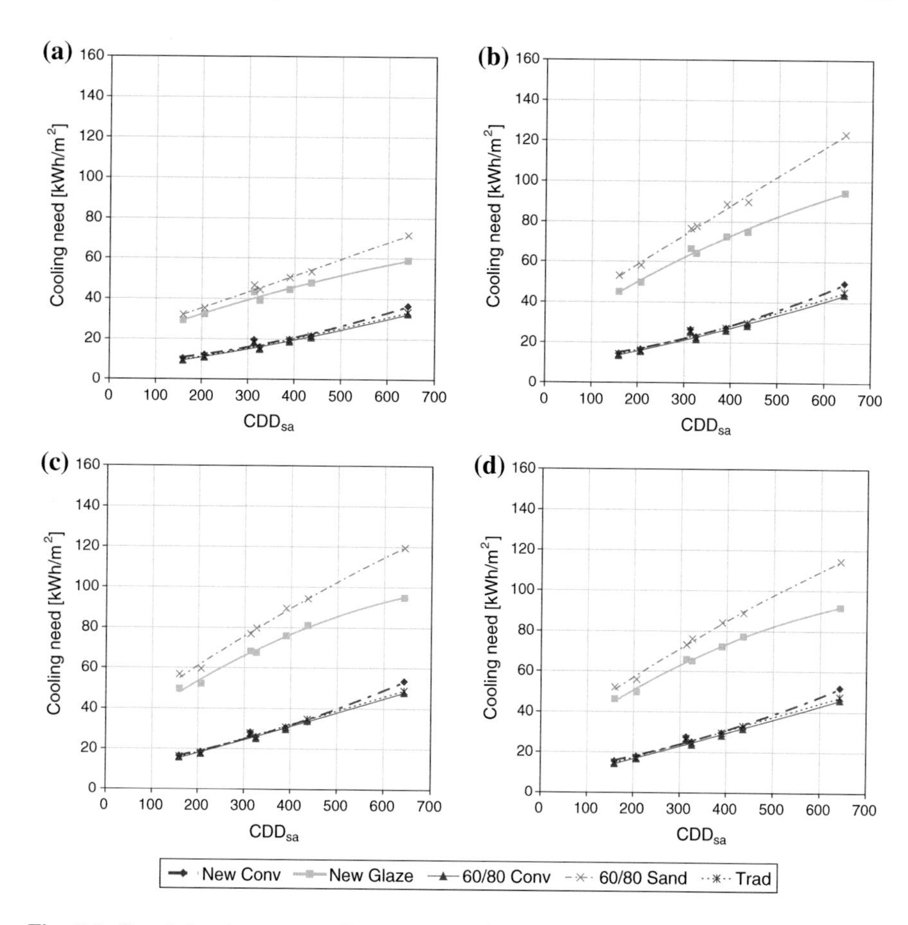

Fig. 7.5 Correlation between cooling need and cooling degree-days, with night ventilation. **a** Orientation North. **b** Orientation South. **c** Orientation East. **d** Orientation West

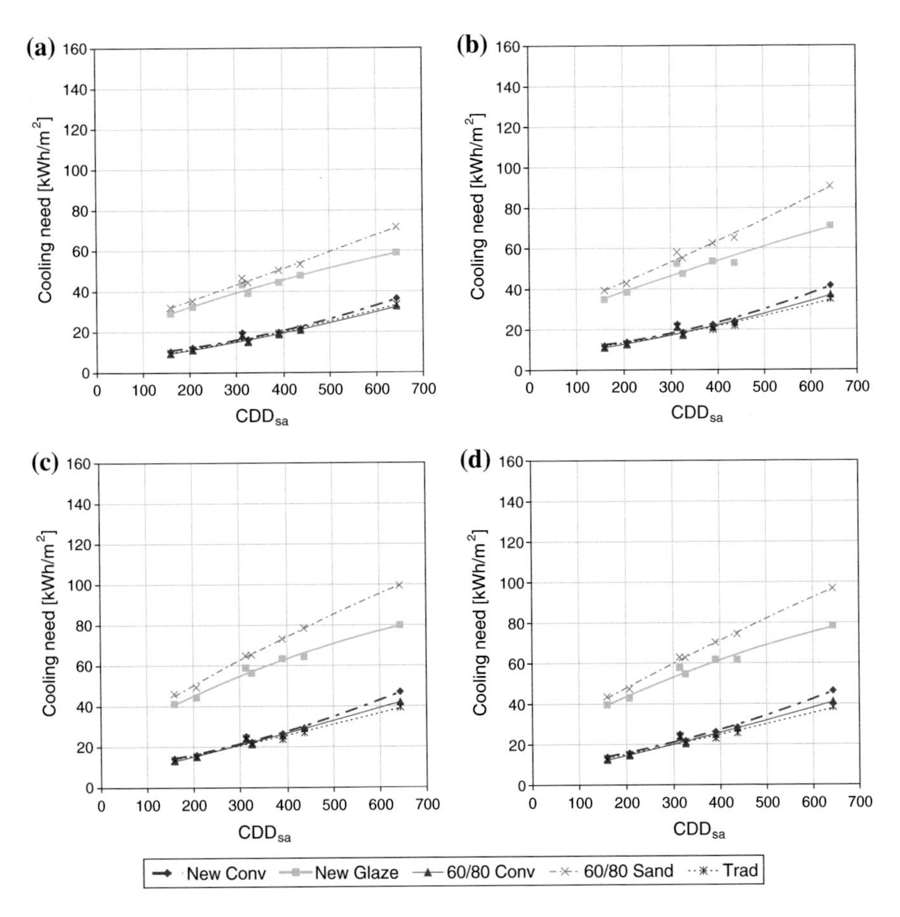

Fig. 7.6 Correlation between cooling need and cooling degree-days, with external shading and night ventilation. **a** Orientation North. **b** Orientation South. **c** Orientation East. **d** Orientation West

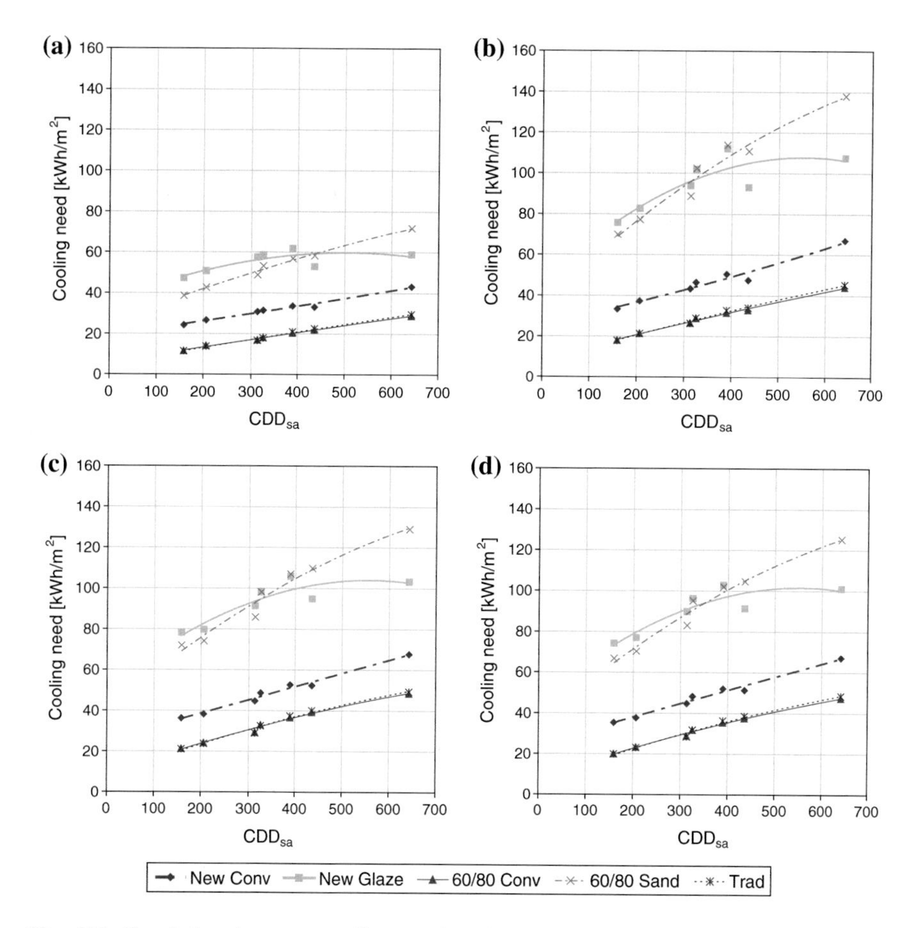

Fig. 7.7 Correlation between cooling need and cooling degree-days. Adaptive set-point. **a** Orientation North. **b** Orientation South. **c** Orientation East. **d** Orientation West

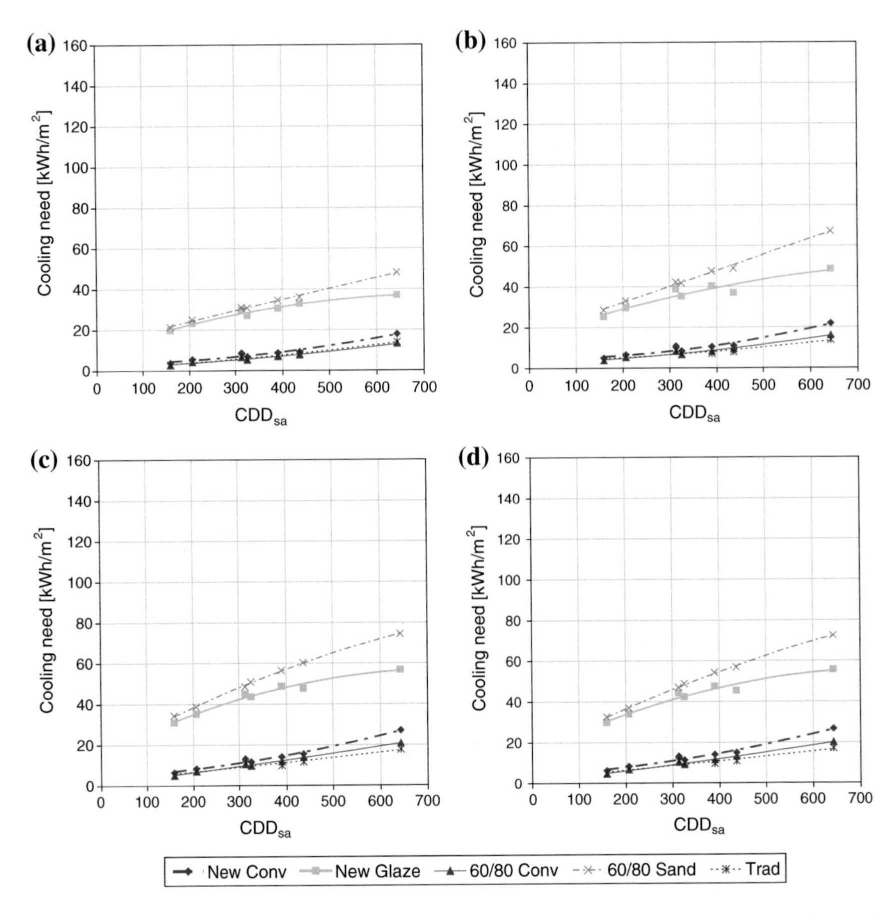

Fig. 7.8 Correlation between cooling need and cooling degree-days, with external shading and night ventilation. Adaptive set-point. **a** Orientation North. **b** Orientation South. **c** Orientation East. **d** Orientation West

7.3 Data Sheets of the Case-Studies Results

In this section, the detailed results for all the considered case-studies are reported in form of data sheets (Figs. 7.9, 7.10, 7.11, 7.12, 7.13, 7.14 and 7.15) and can be directly used for assessing the energy needs of buildings in the same, or comparable, climatic locations.

MILANO

Climate

From D.P.R. 412 (1993)		From Trnsys TRY	
Official Heating Degree-Days	2404	HDD_{sa}	2529
Official Climatic Zone	E	CDD_{sa}	159

Thermal need results

Heating need [kWh/m²]: 0 5 10 20 30 40 50 60

Cooling need [kWh/m²]: 0 10 20 30 60 90 120 150

Change Range [%]: -60 -40 -20 -10 10 20 40 60

New conventional

		Conventional N	S	E	W	Adaptive N	S	E	W	Need change N	S	E	W
HEATING		31.98	24.84	30.28	30.73								
COOLING No Night Vent	No Shading	31.16	39.79	42.82	42.01	24.17	33.35	36.26	35.40	-22%	-16%	-15%	-16%
	Shading	31.16	34.99	38.71	38.63	24.17	28.16	31.92	31.87	-22%	-20%	-18%	-17%
Night Vent	No Shading	10.47	14.53	16.50	15.53	3.89	6.96	8.46	7.82	-63%	-52%	-49%	-50%
	Shading	10.47	12.06	14.15	13.76	3.89	5.03	6.54	6.33	-63%	-58%	-54%	-54%

New glazed

		Conventional N	S	E	W	Adaptive N	S	E	W	Need change N	S	E	W
HEATING		41.10	28.73	37.88	38.96								
COOLING No Night Vent	No Shading	55.09	80.85	85.25	81.38	47.24	75.79	78.23	74.28	-14%	-6%	-8%	-9%
	Shading	55.09	65.67	73.96	72.60	47.24	59.22	66.67	65.19	-14%	-10%	-10%	-10%
Night Vent	No Shading	29.23	44.94	49.50	46.10	19.83	36.47	39.38	36.59	-32%	-19%	-20%	-21%
	Shading	29.23	34.84	41.15	39.50	19.83	25.31	31.04	29.81	-32%	-27%	-25%	-25%

60/80 conventional

		Conventional N	S	E	W	Adaptive N	S	E	W	Need change N	S	E	W
HEATING		73.25	60.78	68.48	69.84								
COOLING No Night Vent	No Shading	20.79	27.81	31.33	29.83	11.63	17.96	21.32	20.02	-44%	-35%	-32%	-33%
	Shading	20.79	23.76	27.42	26.81	11.63	14.17	17.63	17.14	-44%	-40%	-36%	-36%
Night Vent	No Shading	9.27	13.46	15.62	14.20	2.73	5.45	6.87	6.02	-71%	-60%	-56%	-58%
	Shading	9.27	10.95	13.09	12.40	2.73	3.75	5.03	4.64	-71%	-66%	-62%	-63%

60/80 sandwich largely glazed

		Conventional N	S	E	W	Adaptive N	S	E	W	Need change N	S	E	W
HEATING		65.66	47.89	59.51	61.92								
COOLING No Night Vent	No Shading	49.25	77.60	81.97	76.66	38.57	69.88	71.89	66.67	-22%	-10%	-12%	-13%
	Shading	49.25	59.83	68.84	66.47	38.57	50.30	58.41	56.14	-22%	-16%	-15%	-16%
Night Vent	No Shading	31.99	52.91	56.69	51.94	21.30	43.66	44.87	41.10	-33%	-17%	-21%	-21%
	Shading	31.99	39.26	46.05	43.50	21.30	28.50	34.29	32.46	-33%	-27%	-26%	-25%

Traditional

		Conventional N	S	E	W	Adaptive N	S	E	W	Need change N	S	E	W
HEATING		73.99	61.21	69.05	70.43								
COOLING No Night Vent	No Shading	20.78	27.93	31.22	29.89	11.47	17.94	21.06	19.92	-45%	-36%	-33%	-33%
	Shading	20.78	23.91	27.38	26.86	11.47	14.16	17.43	17.03	-45%	-41%	-36%	-37%
Night Vent	No Shading	9.71	14.19	16.18	14.96	2.85	5.88	7.15	6.51	-71%	-59%	-56%	-56%
	Shading	9.71	11.57	13.62	13.09	2.85	4.03	5.26	5.03	-71%	-65%	-61%	-62%

Fig. 7.9 Summary of the results for the case-studies in Milano

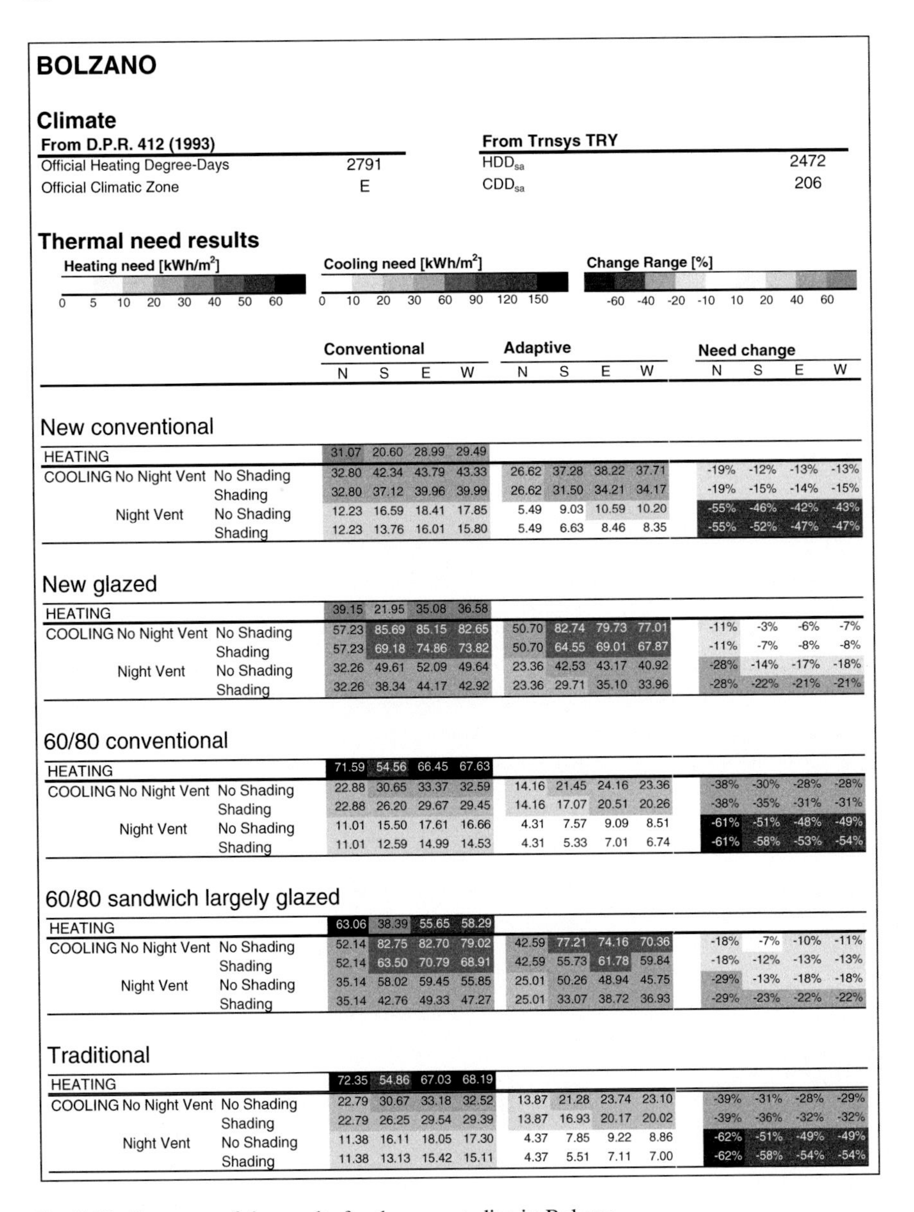

BOLZANO

Climate

From D.P.R. 412 (1993)	
Official Heating Degree-Days	2791
Official Climatic Zone	E

From Trnsys TRY	
HDD_{sa}	2472
CDD_{sa}	206

Thermal need results

Heating need [kWh/m²]: 0 5 10 20 30 40 50 60

Cooling need [kWh/m²]: 0 10 20 30 60 90 120 150

Change Range [%]: -60 -40 -20 -10 10 20 40 60

New conventional

		Conventional N	Conventional S	Conventional E	Conventional W	Adaptive N	Adaptive S	Adaptive E	Adaptive W	Need change N	Need change S	Need change E	Need change W
HEATING		31.07	20.60	28.99	29.49								
COOLING No Night Vent	No Shading	32.80	42.34	43.79	43.33	26.62	37.28	38.22	37.71	-19%	-12%	-13%	-13%
	Shading	32.80	37.12	39.96	39.99	26.62	31.50	34.21	34.17	-19%	-15%	-14%	-15%
Night Vent	No Shading	12.23	16.59	18.41	17.85	5.49	9.03	10.59	10.20	-55%	-46%	-42%	-43%
	Shading	12.23	13.76	16.01	15.80	5.49	6.63	8.46	8.35	-55%	-52%	-47%	-47%

New glazed

		Conventional N	Conventional S	Conventional E	Conventional W	Adaptive N	Adaptive S	Adaptive E	Adaptive W	Need change N	Need change S	Need change E	Need change W
HEATING		39.15	21.95	35.08	36.58								
COOLING No Night Vent	No Shading	57.23	85.69	85.15	82.65	50.70	82.74	79.73	77.01	-11%	-3%	-6%	-7%
	Shading	57.23	69.18	74.86	73.82	50.70	64.55	69.01	67.87	-11%	-7%	-8%	-8%
Night Vent	No Shading	32.26	49.61	52.09	49.64	23.36	42.53	43.17	40.92	-28%	-14%	-17%	-18%
	Shading	32.26	38.34	44.17	42.92	23.36	29.71	35.10	33.96	-28%	-22%	-21%	-21%

60/80 conventional

		Conventional N	Conventional S	Conventional E	Conventional W	Adaptive N	Adaptive S	Adaptive E	Adaptive W	Need change N	Need change S	Need change E	Need change W
HEATING		71.59	54.56	66.45	67.63								
COOLING No Night Vent	No Shading	22.88	30.65	33.37	32.59	14.16	21.45	24.16	23.36	-38%	-30%	-28%	-28%
	Shading	22.88	26.20	29.67	29.45	14.16	17.07	20.51	20.26	-38%	-35%	-31%	-31%
Night Vent	No Shading	11.01	15.50	17.61	16.66	4.31	7.57	9.09	8.51	-61%	-51%	-48%	-49%
	Shading	11.01	12.59	14.99	14.53	4.31	5.33	7.01	6.74	-61%	-58%	-53%	-54%

60/80 sandwich largely glazed

		Conventional N	Conventional S	Conventional E	Conventional W	Adaptive N	Adaptive S	Adaptive E	Adaptive W	Need change N	Need change S	Need change E	Need change W
HEATING		63.06	38.39	55.65	58.29								
COOLING No Night Vent	No Shading	52.14	82.75	82.70	79.02	42.59	77.21	74.16	70.36	-18%	-7%	-10%	-11%
	Shading	52.14	63.50	70.79	68.91	42.59	55.73	61.78	59.84	-18%	-12%	-13%	-13%
Night Vent	No Shading	35.14	58.02	59.45	55.85	25.01	50.26	48.94	45.75	-29%	-13%	-18%	-18%
	Shading	35.14	42.76	49.33	47.27	25.01	33.07	38.72	36.93	-29%	-23%	-22%	-22%

Traditional

		Conventional N	Conventional S	Conventional E	Conventional W	Adaptive N	Adaptive S	Adaptive E	Adaptive W	Need change N	Need change S	Need change E	Need change W
HEATING		72.35	54.86	67.03	68.19								
COOLING No Night Vent	No Shading	22.79	30.67	33.18	32.52	13.87	21.28	23.74	23.10	-39%	-31%	-28%	-29%
	Shading	22.79	26.25	29.54	29.39	13.87	16.93	20.17	20.02	-39%	-36%	-32%	-32%
Night Vent	No Shading	11.38	16.11	18.05	17.30	4.37	7.85	9.22	8.86	-62%	-51%	-49%	-49%
	Shading	11.38	13.13	15.42	15.11	4.37	5.51	7.11	7.00	-62%	-58%	-54%	-54%

Fig. 7.10 Summary of the results for the case-studies in Bolzano

TRIESTE

Climate

From D.P.R. 412 (1993)		From Trnsys TRY	
Official Heating Degree-Days	2102	HDD_{sa}	1750
Official Climatic Zone	E	CDD_{sa}	313

Thermal need results

Heating need [kWh/m²]: 0 5 10 20 30 40 50 60

Cooling need [kWh/m²]: 0 10 20 30 60 90 120 150

Change Range [%]: -60 -40 -20 -10 10 20 40 60

New conventional

			Conventional N	Conventional S	Conventional E	Conventional W	Adaptive N	Adaptive S	Adaptive E	Adaptive W	Need change N	Need change S	Need change E	Need change W
HEATING			15.65	8.14	13.99	14.37								
COOLING	No Night Vent	No Shading	40.20	52.12	53.85	53.97	30.75	43.37	44.62	44.78	-24%	-17%	-17%	-17%
		Shading	40.20	46.01	49.45	49.92	30.75	36.73	40.13	40.64	-24%	-20%	-19%	-19%
	Night Vent	No Shading	19.60	26.52	28.14	27.55	8.76	14.23	15.85	15.55	-55%	-46%	-44%	-44%
		Shading	19.60	22.52	25.06	24.95	8.76	10.84	13.17	13.25	-55%	-52%	-47%	-47%

New glazed

			Conventional N	Conventional S	Conventional E	Conventional W	Adaptive N	Adaptive S	Adaptive E	Adaptive W	Need change N	Need change S	Need change E	Need change W
HEATING			19.06	7.41	15.96	17.02								
COOLING	No Night Vent	No Shading	68.36	101.82	101.36	100.08	57.48	93.93	91.25	89.92	-16%	-8%	-10%	-10%
		Shading	68.36	83.04	89.73	89.79	57.48	73.55	79.39	79.37	-16%	-11%	-12%	-12%
	Night Vent	No Shading	43.21	66.45	68.23	65.97	29.75	53.97	54.28	52.37	-31%	-19%	-20%	-21%
		Shading	43.21	52.42	58.79	57.78	29.75	38.81	44.88	44.10	-31%	-26%	-24%	-24%

60/80 conventional

			Conventional N	Conventional S	Conventional E	Conventional W	Adaptive N	Adaptive S	Adaptive E	Adaptive W	Need change N	Need change S	Need change E	Need change W
HEATING			45.13	31.05	40.75	41.76								
COOLING	No Night Vent	No Shading	29.74	40.53	43.04	42.32	16.89	26.52	29.30	28.72	-43%	-35%	-32%	-32%
		Shading	29.74	34.74	38.59	38.52	16.89	21.03	25.05	25.08	-43%	-39%	-35%	-35%
	Night Vent	No Shading	18.06	25.11	27.12	26.00	6.46	11.31	13.08	12.52	-64%	-55%	-52%	-52%
		Shading	18.06	20.93	23.81	23.34	6.46	8.21	10.49	10.37	-64%	-61%	-56%	-56%

60/80 sandwich largely glazed

			Conventional N	Conventional S	Conventional E	Conventional W	Adaptive N	Adaptive S	Adaptive E	Adaptive W	Need change N	Need change S	Need change E	Need change W
HEATING			36.83	18.27	30.69	32.87								
COOLING	No Night Vent	No Shading	63.39	100.80	100.19	97.38	48.78	88.87	85.91	83.13	-23%	-12%	-14%	-15%
		Shading	63.39	78.51	86.61	85.39	48.78	64.78	72.02	70.88	-23%	-17%	-17%	-17%
	Night Vent	No Shading	46.52	76.56	76.76	73.20	30.95	62.17	60.13	57.25	-33%	-19%	-22%	-22%
		Shading	46.52	57.94	64.80	62.89	30.95	42.17	48.34	46.88	-33%	-27%	-25%	-25%

Traditional

			Conventional N	Conventional S	Conventional E	Conventional W	Adaptive N	Adaptive S	Adaptive E	Adaptive W	Need change N	Need change S	Need change E	Need change W
HEATING			45.72	31.30	41.19	42.19								
COOLING	No Night Vent	No Shading	29.68	40.55	42.84	42.25	16.69	26.37	28.93	28.47	-44%	-35%	-32%	-33%
		Shading	29.68	34.81	38.45	38.47	16.69	20.93	24.75	24.87	-44%	-40%	-36%	-35%
	Night Vent	No Shading	18.54	25.95	27.69	26.87	6.66	11.87	13.43	13.13	-64%	-54%	-52%	-51%
		Shading	18.54	21.67	24.37	24.15	6.66	8.62	10.79	10.88	-64%	-60%	-56%	-55%

Fig. 7.11 Summary of the results for the case-studies in Trieste

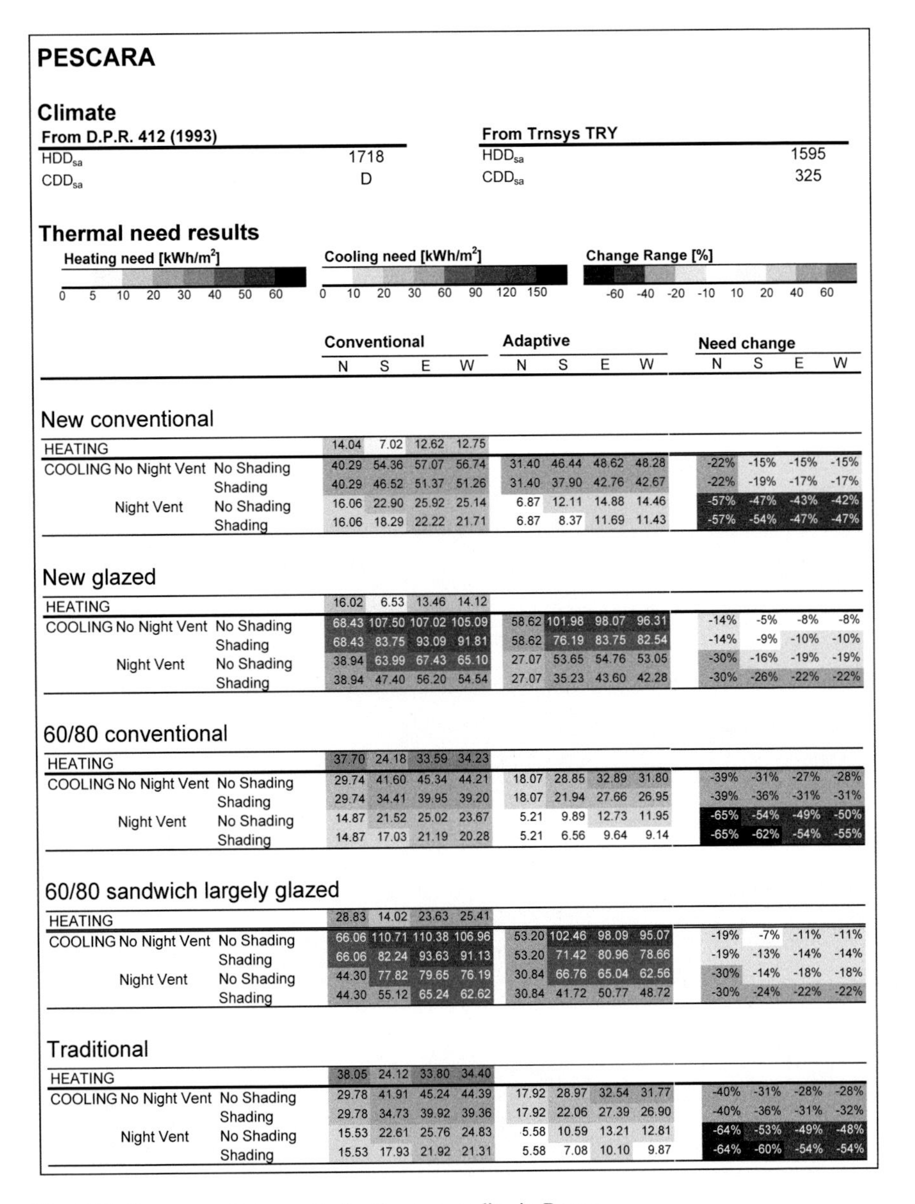

PESCARA

Climate

From D.P.R. 412 (1993)		From Trnsys TRY	
HDD_{sa}	1718	HDD_{sa}	1595
CDD_{sa}	D	CDD_{sa}	325

Thermal need results

Heating need [kWh/m²]: 0 5 10 20 30 40 50 60

Cooling need [kWh/m²]: 0 10 20 30 60 90 120 150

Change Range [%]: -60 -40 -20 -10 10 20 40 60

			Conventional N	Conventional S	Conventional E	Conventional W	Adaptive N	Adaptive S	Adaptive E	Adaptive W	Need change N	Need change S	Need change E	Need change W
New conventional														
HEATING			14.04	7.02	12.62	12.75								
COOLING	No Night Vent	No Shading	40.29	54.36	57.07	56.74	31.40	46.44	48.62	48.28	-22%	-15%	-15%	-15%
		Shading	40.29	46.52	51.37	51.26	31.40	37.90	42.76	42.67	-22%	-19%	-17%	-17%
	Night Vent	No Shading	16.06	22.90	25.92	25.14	6.87	12.11	14.88	14.46	-57%	-47%	-43%	-42%
		Shading	16.06	18.29	22.22	21.71	6.87	8.37	11.69	11.43	-57%	-54%	-47%	-47%
New glazed														
HEATING			16.02	6.53	13.46	14.12								
COOLING	No Night Vent	No Shading	68.43	107.50	107.02	105.09	58.62	101.98	98.07	96.31	-14%	-5%	-8%	-8%
		Shading	68.43	83.75	93.09	91.81	58.62	76.19	83.75	82.54	-14%	-9%	-10%	-10%
	Night Vent	No Shading	38.94	63.99	67.43	65.10	27.07	53.65	54.76	53.05	-30%	-16%	-19%	-19%
		Shading	38.94	47.40	56.20	54.54	27.07	35.23	43.60	42.28	-30%	-26%	-22%	-22%
60/80 conventional														
HEATING			37.70	24.18	33.59	34.23								
COOLING	No Night Vent	No Shading	29.74	41.60	45.34	44.21	18.07	28.85	32.89	31.80	-39%	-31%	-27%	-28%
		Shading	29.74	34.41	39.95	39.20	18.07	21.94	27.66	26.95	-39%	-36%	-31%	-31%
	Night Vent	No Shading	14.87	21.52	25.02	23.67	5.21	9.89	12.73	11.95	-65%	-54%	-49%	-50%
		Shading	14.87	17.03	21.19	20.28	5.21	6.56	9.64	9.14	-65%	-62%	-54%	-55%
60/80 sandwich largely glazed														
HEATING			28.83	14.02	23.63	25.41								
COOLING	No Night Vent	No Shading	66.06	110.71	110.38	106.96	53.20	102.46	98.09	95.07	-19%	-7%	-11%	-11%
		Shading	66.06	82.24	93.63	91.13	53.20	71.42	80.96	78.66	-19%	-13%	-14%	-14%
	Night Vent	No Shading	44.30	77.82	79.65	76.19	30.84	66.76	65.04	62.56	-30%	-14%	-18%	-18%
		Shading	44.30	55.12	65.24	62.62	30.84	41.72	50.77	48.72	-30%	-24%	-22%	-22%
Traditional														
HEATING			38.05	24.12	33.80	34.40								
COOLING	No Night Vent	No Shading	29.78	41.91	45.24	44.39	17.92	28.97	32.54	31.77	-40%	-31%	-28%	-28%
		Shading	29.78	34.73	39.92	39.36	17.92	22.06	27.39	26.90	-40%	-36%	-31%	-32%
	Night Vent	No Shading	15.53	22.61	25.76	24.83	5.58	10.59	13.21	12.81	-64%	-53%	-49%	-48%
		Shading	15.53	17.93	21.92	21.31	5.58	7.08	10.10	9.87	-64%	-60%	-54%	-54%

Fig. 7.12 Summary of the results for the case-studies in Pescara

ROMA

Climate

From D.P.R. 412 (1993)		From Trnsys TRY	
Official Heating Degree-Days	1415	HDD_{sa}	1397
Official Climatic Zone	D	CDD_{sa}	390

Thermal need results

Heating need [kWh/m²]: 0 5 10 20 30 40 50 60

Cooling need [kWh/m²]: 0 10 20 30 60 90 120 150

Change Range [%]: -60 -40 -20 -10 10 20 40 60

New conventional

		Conventional N	Conventional S	Conventional E	Conventional W	Adaptive N	Adaptive S	Adaptive E	Adaptive W	Need change N	Need change S	Need change E	Need change W
HEATING		10.57	3.71	8.96	9.24								
COOLING No Night Vent	No Shading	43.89	58.89	62.18	61.51	33.70	50.46	52.64	52.00	-23%	-14%	-15%	-15%
	Shading	43.89	50.20	55.80	55.95	33.70	40.78	45.96	46.18	-23%	-19%	-18%	-17%
Night Vent	No Shading	19.80	27.15	30.41	29.62	8.77	14.55	17.57	17.12	-56%	-46%	-42%	-42%
	Shading	19.80	22.38	26.38	26.19	8.77	10.52	14.04	14.02	-56%	-53%	-47%	-46%

New glazed

		Conventional N	Conventional S	Conventional E	Conventional W	Adaptive N	Adaptive S	Adaptive E	Adaptive W	Need change N	Need change S	Need change E	Need change W
HEATING		11.81	2.79	8.96	9.75								
COOLING No Night Vent	No Shading	73.17	118.56	116.26	113.15	61.80	112.10	105.82	102.86	-16%	-5%	-9%	-9%
	Shading	73.17	91.58	100.20	99.46	61.80	83.21	89.34	88.68	-16%	-9%	-11%	-11%
Night Vent	No Shading	44.47	72.68	75.93	72.34	30.53	61.69	61.53	58.47	-31%	-15%	-19%	-19%
	Shading	44.47	53.58	63.33	61.64	30.53	40.27	48.73	47.50	-31%	-25%	-23%	-23%

60/80 conventional

		Conventional N	Conventional S	Conventional E	Conventional W	Adaptive N	Adaptive S	Adaptive E	Adaptive W	Need change N	Need change S	Need change E	Need change W
HEATING		31.61	16.92	27.05	27.79								
COOLING No Night Vent	No Shading	33.78	45.83	50.84	49.46	20.35	31.56	36.54	35.33	-40%	-31%	-28%	-29%
	Shading	33.78	38.73	44.81	44.47	20.35	24.50	30.71	30.43	-40%	-37%	-31%	-32%
Night Vent	No Shading	18.69	25.88	29.63	28.27	6.90	12.02	15.19	14.33	-63%	-54%	-49%	-49%
	Shading	18.69	21.12	25.44	24.86	6.90	8.34	11.78	11.49	-63%	-60%	-54%	-54%

60/80 sandwich largely glazed

		Conventional N	Conventional S	Conventional E	Conventional W	Adaptive N	Adaptive S	Adaptive E	Adaptive W	Need change N	Need change S	Need change E	Need change W
HEATING		22.95	8.05	17.14	19.20								
COOLING No Night Vent	No Shading	71.69	123.28	121.03	115.71	56.84	113.80	106.93	101.90	-21%	-8%	-12%	-12%
	Shading	71.69	90.75	101.70	99.44	56.84	78.86	87.17	85.10	-21%	-13%	-14%	-14%
Night Vent	No Shading	50.35	88.68	89.46	84.11	34.46	76.62	72.80	68.28	-32%	-14%	-19%	-19%
	Shading	50.35	62.38	73.15	70.35	34.46	47.67	56.33	54.17	-32%	-24%	-23%	-23%

Traditional

		Conventional N	Conventional S	Conventional E	Conventional W	Adaptive N	Adaptive S	Adaptive E	Adaptive W	Need change N	Need change S	Need change E	Need change W
HEATING		30.16	15.74	25.67	26.38								
COOLING No Night Vent	No Shading	34.44	46.92	51.62	50.42	20.99	32.78	37.37	36.36	-39%	-30%	-28%	-28%
	Shading	34.44	39.66	45.55	45.36	20.99	25.48	31.47	31.35	-39%	-36%	-31%	-31%
Night Vent	No Shading	19.50	27.17	30.63	29.60	7.53	13.11	16.08	15.55	-61%	-52%	-47%	-47%
	Shading	19.50	19.91	23.60	22.99	7.53	7.19	9.88	9.68	-61%	-64%	-58%	-58%

Fig. 7.13 Summary of the results for the case-studies in Roma

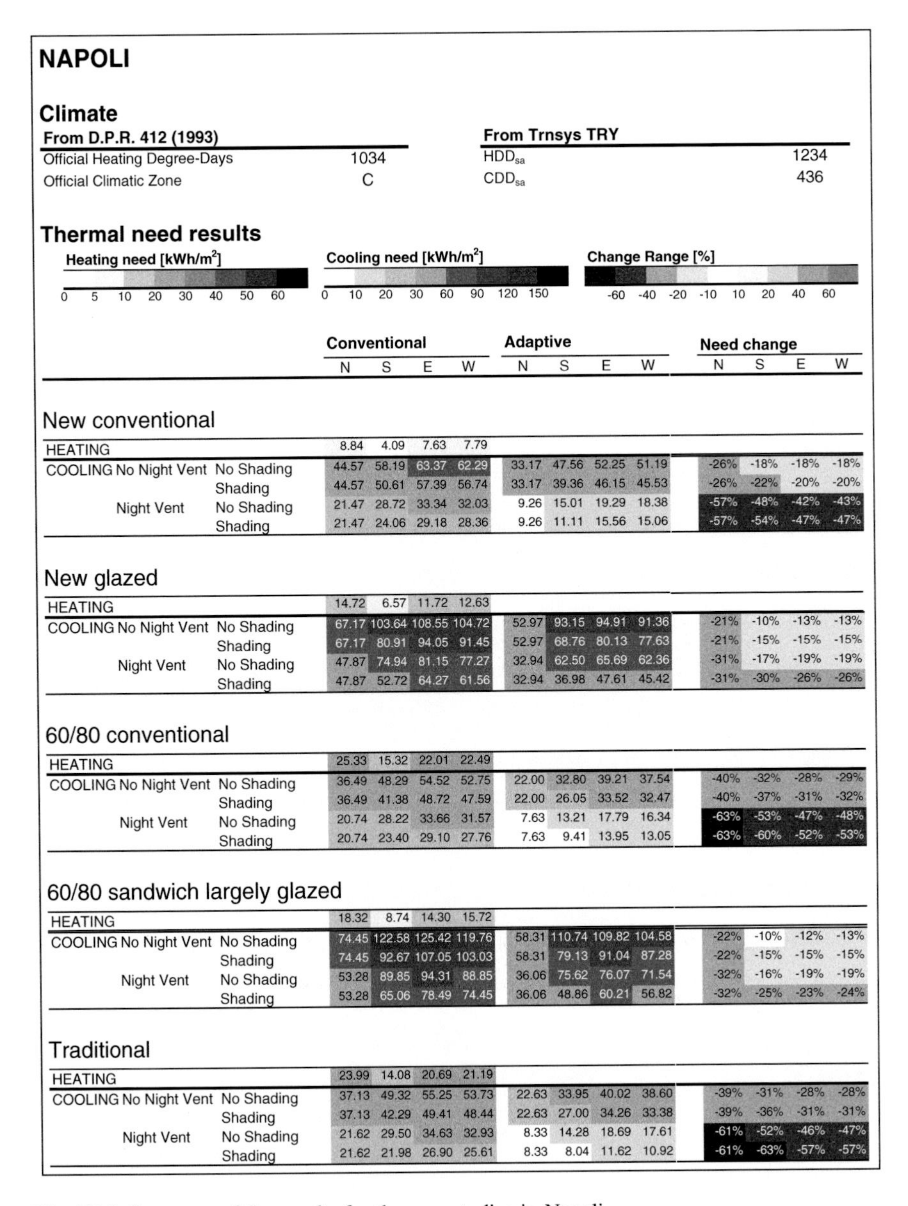

NAPOLI

Climate

From D.P.R. 412 (1993)		From Trnsys TRY	
Official Heating Degree-Days	1034	HDD_{sa}	1234
Official Climatic Zone	C	CDD_{sa}	436

Thermal need results

New conventional

		Conventional N	S	E	W	Adaptive N	S	E	W	Need change N	S	E	W
HEATING		8.84	4.09	7.63	7.79								
COOLING No Night Vent	No Shading	44.57	58.19	63.37	62.29	33.17	47.56	52.25	51.19	-26%	-18%	-18%	-18%
	Shading	44.57	50.61	57.39	56.74	33.17	39.36	46.15	45.53	-26%	-22%	-20%	-20%
Night Vent	No Shading	21.47	28.72	33.34	32.03	9.26	15.01	19.29	18.38	-57%	-48%	-42%	-43%
	Shading	21.47	24.06	29.18	28.36	9.26	11.11	15.56	15.06	-57%	-54%	-47%	-47%

New glazed

		Conventional N	S	E	W	Adaptive N	S	E	W	Need change N	S	E	W
HEATING		14.72	6.57	11.72	12.63								
COOLING No Night Vent	No Shading	67.17	103.64	108.55	104.72	52.97	93.15	94.91	91.36	-21%	-10%	-13%	-13%
	Shading	67.17	80.91	94.05	91.45	52.97	68.76	80.13	77.63	-21%	-15%	-15%	-15%
Night Vent	No Shading	47.87	74.94	81.15	77.27	32.94	62.50	65.69	62.36	-31%	-17%	-19%	-19%
	Shading	47.87	52.72	64.27	61.56	32.94	36.98	47.61	45.42	-31%	-30%	-26%	-26%

60/80 conventional

		Conventional N	S	E	W	Adaptive N	S	E	W	Need change N	S	E	W
HEATING		25.33	15.32	22.01	22.49								
COOLING No Night Vent	No Shading	36.49	48.29	54.52	52.75	22.00	32.80	39.21	37.54	-40%	-32%	-28%	-29%
	Shading	36.49	41.38	48.72	47.59	22.00	26.05	33.52	32.47	-40%	-37%	-31%	-32%
Night Vent	No Shading	20.74	28.22	33.66	31.57	7.63	13.21	17.79	16.34	-63%	-53%	-47%	-48%
	Shading	20.74	23.40	29.10	27.76	7.63	9.41	13.95	13.05	-63%	-60%	-52%	-53%

60/80 sandwich largely glazed

		Conventional N	S	E	W	Adaptive N	S	E	W	Need change N	S	E	W
HEATING		18.32	8.74	14.30	15.72								
COOLING No Night Vent	No Shading	74.45	122.58	125.42	119.76	58.31	110.74	109.82	104.58	-22%	-10%	-12%	-13%
	Shading	74.45	92.67	107.05	103.03	58.31	79.13	91.04	87.28	-22%	-15%	-15%	-15%
Night Vent	No Shading	53.28	89.85	94.31	88.85	36.06	75.62	76.07	71.54	-32%	-16%	-19%	-19%
	Shading	53.28	65.06	78.49	74.45	36.06	48.86	60.21	56.82	-32%	-25%	-23%	-24%

Traditional

		Conventional N	S	E	W	Adaptive N	S	E	W	Need change N	S	E	W
HEATING		23.99	14.08	20.69	21.19								
COOLING No Night Vent	No Shading	37.13	49.32	55.25	53.73	22.63	33.95	40.02	38.60	-39%	-31%	-28%	-28%
	Shading	37.13	42.29	49.41	48.44	22.63	27.00	34.26	33.38	-39%	-36%	-31%	-31%
Night Vent	No Shading	21.62	29.50	34.63	32.93	8.33	14.28	18.69	17.61	-61%	-52%	-46%	-47%
	Shading	21.62	21.98	26.90	25.61	8.33	8.04	11.62	10.92	-61%	-63%	-57%	-57%

Fig. 7.14 Summary of the results for the case-studies in Napoli

PALERMO

Climate

From D.P.R. 412 (1993)		From Trnsys TRY	
Official Heating Degree-Days	751	HDD_{sa}	551
Official Climatic Zone	B	CDD_{sa}	643

Thermal need results

Heating need [kWh/m²]: 0 5 10 20 30 40 50 60

Cooling need [kWh/m²]: 0 10 20 30 60 90 120 150

Change Range [%]: -60 -40 -20 -10 10 20 40 60

New conventional

		Conventional N	Conventional S	Conventional E	Conventional W	Adaptive N	Adaptive S	Adaptive E	Adaptive W	Need change N	Need change S	Need change E	Need change W
HEATING		0.65	0.13	0.49	0.50								
COOLING No Night Vent	No Shading	59.69	82.83	84.17	83.64	43.18	66.99	67.64	67.16	-28%	-19%	-20%	-20%
	Shading	59.69	69.81	75.37	75.74	43.18	53.50	58.74	59.15	-28%	-23%	-22%	-22%
Night Vent	No Shading	36.47	49.36	53.46	51.84	17.92	28.82	32.99	31.76	-51%	-42%	-38%	-39%
	Shading	36.47	41.70	46.99	46.34	17.92	21.92	27.07	26.64	-51%	-47%	-42%	-42%

New glazed

		Conventional N	Conventional S	Conventional E	Conventional W	Adaptive N	Adaptive S	Adaptive E	Adaptive W	Need change N	Need change S	Need change E	Need change W
HEATING		2.58	0.48	1.65	1.89								
COOLING No Night Vent	No Shading	79.40	127.04	123.84	121.59	58.87	107.61	103.29	101.10	-26%	-15%	-17%	-17%
	Shading	79.40	98.78	106.31	105.82	58.87	78.88	85.57	85.25	-26%	-20%	-20%	-19%
Night Vent	No Shading	58.91	94.42	94.88	91.70	36.97	72.29	71.34	68.83	-37%	-23%	-25%	-25%
	Shading	58.91	71.02	79.88	78.41	36.97	48.57	56.68	55.64	-37%	-32%	-29%	-29%

60/80 conventional

		Conventional N	Conventional S	Conventional E	Conventional W	Adaptive N	Adaptive S	Adaptive E	Adaptive W	Need change N	Need change S	Need change E	Need change W
HEATING		5.12	1.38	3.71	4.06								
COOLING No Night Vent	No Shading	48.54	64.85	69.06	67.89	28.87	44.05	48.46	47.36	-41%	-32%	-30%	-30%
	Shading	48.54	55.66	61.69	61.46	28.87	35.01	41.33	41.15	-41%	-37%	-33%	-33%
Night Vent	No Shading	32.47	43.64	47.78	45.85	13.20	21.66	25.76	24.33	-59%	-50%	-46%	-47%
	Shading	32.47	37.03	42.01	41.06	13.20	16.08	20.82	20.17	-59%	-57%	-50%	-51%

60/80 sandwich largely glazed

		Conventional N	Conventional S	Conventional E	Conventional W	Adaptive N	Adaptive S	Adaptive E	Adaptive W	Need change N	Need change S	Need change E	Need change W
HEATING		2.46	0.50	1.55	1.80								
COOLING No Night Vent	No Shading	94.09	158.80	151.15	147.28	71.88	137.93	129.03	125.14	-24%	-13%	-15%	-15%
	Shading	94.09	121.85	128.36	127.11	71.88	100.58	106.07	104.91	-24%	-17%	-17%	-17%
Night Vent	No Shading	71.70	123.16	119.40	114.36	48.04	99.92	94.32	89.98	-33%	-19%	-21%	-21%
	Shading	71.70	90.66	99.41	96.83	48.04	67.08	74.49	72.48	-33%	-26%	-25%	-25%

Traditional

		Conventional N	Conventional S	Conventional E	Conventional W	Adaptive N	Adaptive S	Adaptive E	Adaptive W	Need change N	Need change S	Need change E	Need change W
HEATING		4.57	1.17	3.26	3.60								
COOLING No Night Vent	No Shading	49.16	65.98	69.79	68.89	29.57	45.35	49.33	48.50	-40%	-31%	-29%	-30%
	Shading	49.16	56.54	62.38	62.35	29.57	35.99	42.15	42.16	-40%	-36%	-32%	-32%
Night Vent	No Shading	33.43	45.02	48.84	47.35	14.00	22.98	26.76	25.77	-58%	-49%	-45%	-46%
	Shading	33.43	34.97	39.30	38.29	14.00	13.78	17.68	17.07	-58%	-61%	-55%	-55%

Fig. 7.15 Summary of the results for the case-studies in Palermo

References

ASHRAE, *Handbook of Fundamentals* (American Society of Heating, Refrigerating and Air-conditioning Engineers, Atlanta, 2001)

D.P.R. 412, *Regolamento recante norme per la progettazione, l'installazione e la manutenzione degli impianti termici degli edifici, ai fini del contenimento dei consumi di energia, in attuazione dell'art. 4, comma 4 della legge 9 gennaio 1991, n.10. (aggiornata dal D.P. R.551/99)* (Rome, 1993)

S. Ferrari, V. Zanotto, Energy performance of different office building envelopes within the Italian context, in *Proceedings of the 48th AICARR International Conference—Energy refurbishment of existing buildings—which solutions for an integrated system: envelope, plant, controls*, Baveno, 22–23 Sep 2011

S. Ferrari, V. Zanotto, Adaptive comfort: analysis and application of the main indices. Build. Environ. **49**, 25–32 (2012a)

S. Ferrari, V. Zanotto, Office buildings cooling need in the Italian climatic context: assessing the performances of typical envelopes. Energ. Procedia **30**, 1099–1109 (2012b)

S.A. Klein, W.A. Beckman, J.W. Mitchell et al., *TRNSYS—a Transient System Simulation Program User Manual* (The Solar Energy Laboratory—University of Wisconsin, Madison, 2007)

Chapter 8
Buildings Performance Comparison: From Energy Need to Energy Consumption

Abstract Building heating and cooling energy needs result from the thermal balance calculation, and can be determined with a high degree of accuracy through the use of detailed dynamic energy simulation tools. On this basis, starting from a reference building shape, assuming the same time-related input parameters (such as the internal sources of heat, the ventilation and the indoor temperatures) and varying the construction elements, the energy performance of a set of alternative building solutions can be evaluated and compared. However, the related building energy consumption, which is also based on the characteristics of the active air conditioning systems, does not necessarily depend linearly on the calculated heating and cooling needs. A further overall energy performance comparison between the different building solutions should therefore be adjusted accordingly. In this chapter, reference climatization systems are defined and assigned to a set of case-study buildings. Final and primary energy consumption values are then calculated and the differences in the overall energy performance trends, compared to those based on the calculated energy needs, are assessed.

Keywords Overall building energy performance · Building energy need and consumption · Base HVAC systems · HVAC auxiliaries consumption · Building's primary energy

8.1 HVAC Systems and Primary Energy Consumption

HVAC is an acronym meaning "heating, ventilation and air conditioning", and refers to the active climatization system, which is devoted to provide healthy and comfortable indoor environments to the occupants.

Under the energy point of view, HVAC systems can be briefly described according to four main components (generally called systems too):

- generation, which is the part providing the energy that will be used to increase or decrease building indoor temperature;

S. Ferrari and V. Zanotto, *Building Energy Performance Assessment in Southern Europe*, PoliMI SpringerBriefs,
DOI 10.1007/978-3-319-24136-4_8

- distribution, which is the part transmitting energy from the generation machines to the building spaces;
- emission terminals, which constitute the interface of energy exchange between the distribution and the indoor environments, and are usually placed inside each room of the building;
- regulation, which determines the operating mode of the system.

The most common thermal generation systems installed at building level are fuel based heating boilers and electrical based chillers for cooling or electrical based heat pumps for both heating and cooling.

The distribution system is mainly characterized by the heat distribution mean that is usually either water or air. Water is generally a more efficient vector, since its higher specific heat and mass density allow the recourse to small flows, and therefore the use of pipes with reduced sections and pumps with small sizes (Rossi 2002). Air, on the other hand, which is generally distributed through rectangular or circular ducts, allows to combine the heating and cooling function with air conditioning, controlling humidity,[1] freshness and quality of the indoor air (by the means of an air-handling unit).

The emission terminals are the element directly changing the indoor environment conditions and are commonly of two types:

- convective systems, which act on the indoor air temperature by exchanging thermal energy directly with the air mass in the room or through the introduction of a new conditioned (primary) air mass;
- radiant systems, which act on the temperature of one or more surfaces in the room, and therefore change predominantly the indoor mean radiant temperature.

The given requirements for thermal comfort regard the environment operative temperature (EN ISO 15251 2007), which is a linear combination of air temperature and mean radiant temperature that can be approximated, in most cases, to a simple average between the two values (EN ISO 7726 1998): therefore the convective and radiant systems work on the two components of this value, influencing the indoor environmental conditions in very different ways. The regulation of the heating and cooling system, anyway, is usually performed only by the means of air temperature sensors and set-points.[2]

The first emission terminal category refers to local diffusers (e.g. vents, anemostats, etc.) in case of distribution systems air based or to local water-to-air heat exchangers in case of distribution systems water based, driven by natural

[1]The control of air humidity is very important in summer, since the latent component due to the heat stored in the water vapor present in the air can determine the increase of the cooling loads. However, this element is very hard to determine without considering a complex air conditioning system, and it is usually neglected when assessing the building performances in first analysis.

[2]In case of radiant systems this kind of regulation is not properly effective, since the way the climatization system performs is not directly taken into account.

convection (e.g. radiators[3]) or by forced convection (provided with fans, such as fan-coils), while the second category refers to the water distribution systems that locally branch into pipe coils creating radiant surfaces, such as radiant floor, ceiling and wall or radiant strips.

For a proper assessment of the final energy consumption the specific thermal efficiencies of these system components as well as the electricity consumptions of the related auxiliaries (mainly fans and pumps) have to be considered after calculating the building heating and cooling needs. The electricity component, in particular, is very often neglected within the building energy performance assessment procedures, but can in some cases constitute a very important part of the actual overall consumption of the buildings.

Finally, in order to summarize the different energy sources involved in the overall building heating and cooling process, all the final energy consumption values, both fuel and electrical based, have to be converted into primary energy ones. It has to be noted that, concerning the electrical consumption in particular, the corresponding primary energy use can vary significantly depending on the electric generation sources mix of the considered nation.

8.2 Application on Case Studies

Starting from the energy needs calculated for comparing the performances of typical buildings within the Italian context described in the previous Chaps. 6 and 7, the building energy consumptions due to the adoption of a climatization system has been estimated and the change in the overall energy performance trends has been assessed.

Considering the same reference building shape of five storey with rectangular plan (30 m × 12 m), the case-studies were again defined, based on both different envelope characteristics and window percentage, to be representative of likely practices from three main construction ages: newly built, 1960/80 and very old.

For every construction age, "conventional" vertical envelope solutions have been taken into account, with masonry external walls and a window surface equal to 1/8 of the floor area (the rooms and their openings are located on the two main façades). Moreover, alternative more glazed and lighter solutions have been considered for the contemporary and 1960/80 ages.

The resulting building models are five:

- two new buildings (compliant with recent regulations):
  - with conventional façades;
  - with completely glazed façades;

[3]Despite the name, since its surface is very limited the radiative effect on the indoor environment is weak while the convection phenomena, determined by the increased temperature its elements, are predominant.

- two buildings from 1960/1980:
  - with conventional façades;
  - with insulated sandwich walls largely glazed;
- one building which represents the traditional old buildings of the area

The matrix of the simulations was defined also based on seven climatic locations and four main building orientations (detailed description in Chap. 5).

Once the building energy needs and the peak loads (assessing the size of the climatization systems) have been calculated, by the means of the building heat balance computed through TRNSYS simulation software (Klein et al. 2007), the changing in the overall energy performance trends due to the adoption of a climatization system has been evaluated and compared among the different building construction. For implementing this further analysis, the simplified assessment method and reference values describing active systems behaviour suggested by the Italian technical standards to assess the energy performance of buildings have been used (UNI TS 11300-2 2008; UNI TS 11300-3 2010).

8.2.1 Energy Performances Comparison

An example taken from the matrix is here reported, as representative of the entire set. It refers to the building models placed in the Rome, location that can represent the national mean climatic condition, having the main façades oriented towards North-South and in standard conditions (indoor set-point temperatures between 20 and 26 °C, without considering passive cooling strategies). The graphs in Figs. 8.1 and 8.2 report the heating and cooling needs and the size of the climatization systems of the considered building models.

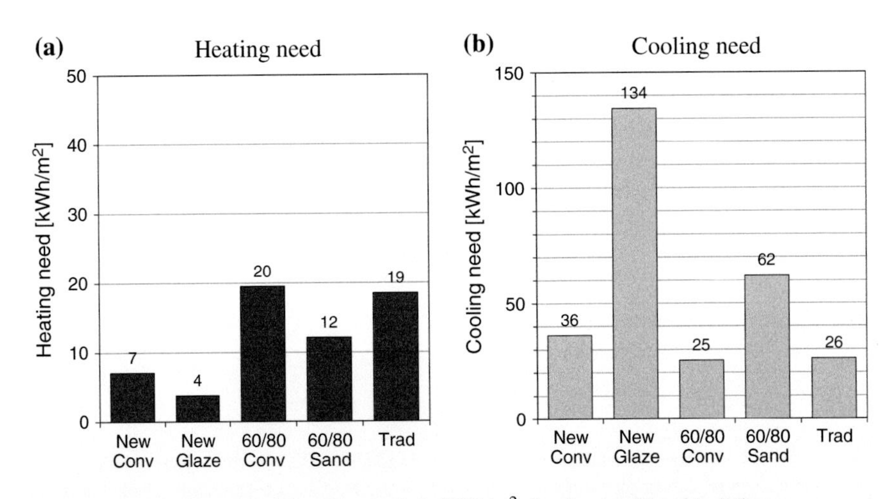

Fig. 8.1 Heating (**a**) and cooling (**b**) needs, in kWh/m^2, for the considered buildings

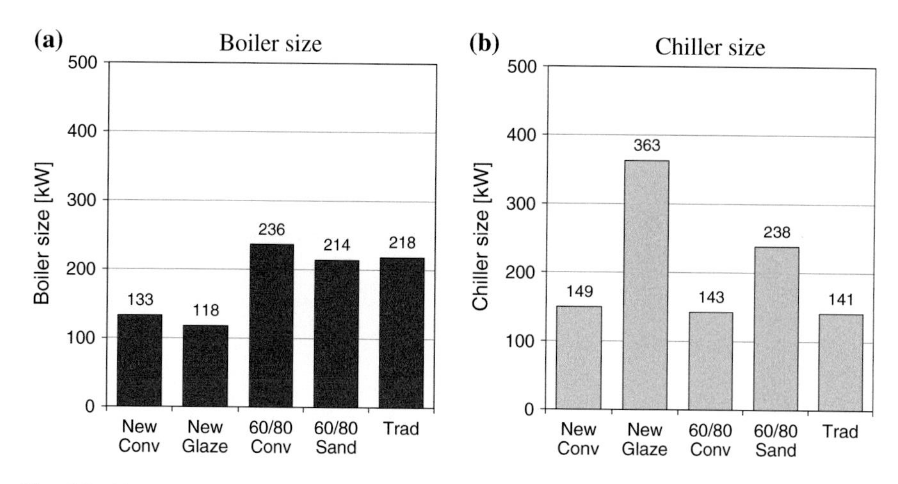

Fig. 8.2 Size, in kW, of the boilers (**a**) and chillers (**b**) needed by the building according to the different envelope constructions

The study aim was to evaluate the energy consumption pattern of common Italian office buildings, and therefore a widespread active system is considered. It is a 4 pipes fan-coil system, providing both heating and cooling, while a gas boiler and an electrical chiller, both centralized, are in charge of the generation part (Alfano et al. 1997; Cammarata 2007; Dall'O' 1999).

The Italian technical standards UNI TS 11300-2 (2008) and UNI TS 11300-3 (2010), which are prescribed by the Italian regulation within the building certification framework, introduce the way all the system components have to be taken into account in the calculation of the primary energy consumption. According to these technical standards, the global efficiency and the electricity consumption of the active system can be determined as combination of the partial efficiency and auxiliary consumption of each single system component: within the standards, general average values are given for the different system typologies which can be found in Italy.

Concerning heating, the generation and distribution systems characteristics have been assigned based on the buildings' ages, according to the reference data reported in UNI TS 11300-2 (2008) (Table 8.1).

Table 8.1 Average characteristics of the considered heating systems

	New	1960/1980	Old
Generation (boiler)	Condensing (η 0.99)	Common (η 0.88)	Common (η 0.88)
Distribution	Horizontal (η 0.99) Variable speed pumps	Vertical (1961–76) (η 0.96) Constant speed pumps	Vertical (before 1961) (η 0.95) Constant speed pumps
Emission	Fan-coils (η 0.95)		
Regulation	Climatic (outdoor sensor) and environmental with on-off switch (η 0.95)		

Table 8.2 Average characteristics of the considered cooling systems

	New	1960/80	Old
Generation (chiller)	Air-water (COP 2)		
Distribution	Horizontal (η 0.99) Variable speed pumps	Vertical (1961–76) (η 0.99) Constant speed pumps	Vertical (before 1961) (η 0.99) Constant speed pumps
Emission	Fan-coils (η 0.98)		
Regulation	Environmental with on-off switch (η 0.94)		

Concerning cooling, it is considered that the distribution net, the emission terminals and the regulation are the same of the heating system ones, assuming the connected reference data prescribed by UNI TS 11300-3 (2010) and reported in Table 8.2.

Differently, always according to the technical standard, the chiller characteristics should be derived by the performance curves provided for specific products: in order to apply general values, in this study a conservative seasonal COP of 2 has been considered. The electric energy consumption due to the chiller auxiliaries have been calculated on a monthly basis according to size, number of working hours and load factor of the system, as suggested in the standard.

Hence, the final energy consumption due to space heating and cooling was derived according to the heating and cooling need computed through the TRNSYS (Klein et al. 2007) dynamic simulations.

For assessing the electrical energy consumption due to the auxiliaries of the emission and the distribution systems (fan-coil fans and hydronic system circulation pumps), the following data have been assumed for each one of the 60 offices in the building:

- average air flow of 300 m^3/h;
- water flow of 900 dm^3/h during the heating season and of 700 dm^3/h during the cooling season.

Hence, the distribution and emission electrical energy consumption of the auxiliaries for the whole building is calculated as a summation of 60 single offices consumptions.

In order to compare the overall energy consumption of the different buildings, the final energy values were converted in primary energy consumption, as suggested within the technical standard UNI TS 11300-2 (2008):

- natural gas has a conversion factor of 1;
- electrical energy is converted in equivalent oil tons through the value of 0.187×10^{-3} Tep/kWh_{el} (Delibera EEN 3 2008) and then transformed in kWh of primary energy through the conversion factor of 11.86×10^3 kWh_{prim}/Tep prescribed by UNI TS 11300-2 (2008), which corresponds to an overall conversion efficiency of 0.45.

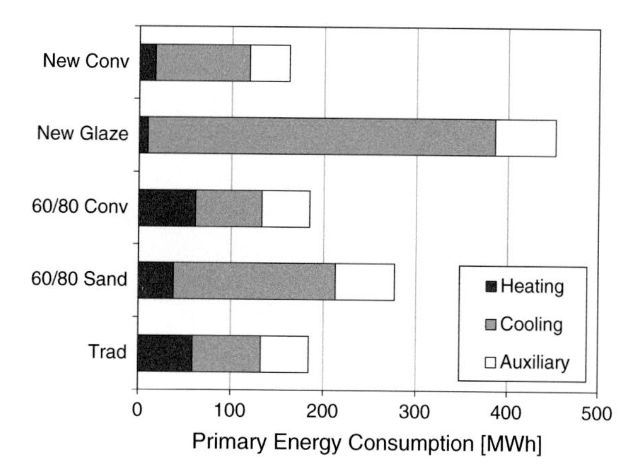

Fig. 8.3 Primary energy consumption, in MWh, for the analysed buildings connected to climatization

Table 8.3 Calculated energy consumption, in kWh/m^2, for the analysed buildings

	Heating (gas) [kWh/m^2]	Cooling (electricity) [kWh/m^2]	Auxiliaries (electricity) [kWh/m^2]	Global primary energy [kWh/m^2]
New conventional	7.83	20.29	8.62	71.96
New glazed	4.19	75.45	13.38	201.19
60/80 conventional	27.53	14.18	10.37	81.99
60/80 sandwich	17.12	34.86	12.93	123.11
Traditional	26.38	14.62	10.38	81.83

The results in Fig. 8.3 show that the primary energy connected to summer cooling is higher than the one due to winter heating in every one of the considered buildings. This is true also in the cases "Traditional" and "60/80 Conventional" in which the final energy heating component is higher than the cooling one (Table 8.3).

By comparing the primary energy results in Fig. 8.3 with the energy need reported in Fig. 7.1, it has to be noted that the solution "New Glazed", despite its best heating energy need performance, has the worst cooling need performance that determines the worst case in terms of overall primary energy consumption. Moreover, the solutions "Traditional" and "60/80 Conventional" perform the worst in terms of heating need and the best in terms of cooling needs but the overall results, in terms of primary energy, is quite similar to the "New Conventional" solution.

These aspects highlight the importance of a proper detailed evaluation of the energy performance of the buildings, with particular reference to the summer behaviour and especially in southern Europe climatic context, because of the close relation between the building construction characteristics and the effect on the overall primary energy consumption.

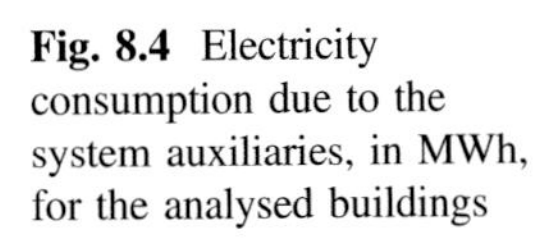

Fig. 8.4 Electricity consumption due to the system auxiliaries, in MWh, for the analysed buildings

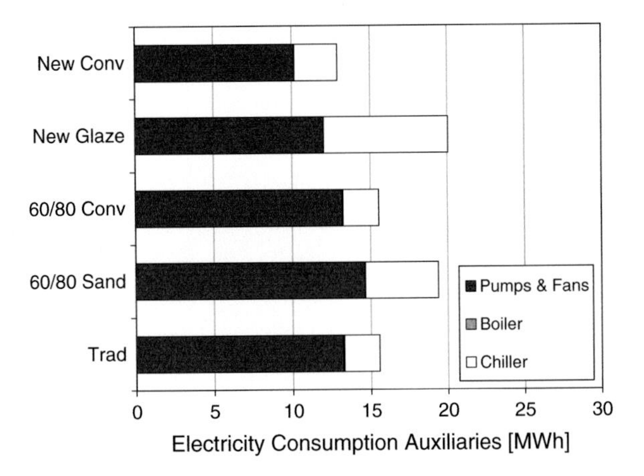

With this respect, finally, the electrical energy use due to the system auxiliaries has been analysed in detail.

The results in Fig. 8.4 show that, among all the considered buildings, the amount of energy consumption related to the auxiliaries of the distribution (pumps) and emission (fans) components is of comparable entity, the one of the heating generation system is extremely low (unreadable in the graph), while the one of the cooling generation system shows big differences (due to the fans in the chiller) and is much more relevant in case of the new glazed building.

References

G. Alfano, M. Filippi, E. Sacchi, *Impianti per la climatizzazione dell'edilizia: dal progetto al collaudo* (Masson, Milano, 1997)

G. Cammarata, *Impianti termotecnica—Volume secondo* (Facoltà di Ingegneria—Università di Catania, 2007)

G. Dall'O', *Architettura e impianti—Tecnologie dei sistemi impiantistici negli uffici* (CittàStudi Edizioni, Torino, 1999)

Delibera EEN 3, *Aggiornamento del fattore di conversione dei kWh in tonnellate equivalenti di petrolio connesso al meccanismo dei titoli di efficienza energetica* (Autorità per l'Energia Elettrica e il Gas, Roma, 2008)

EN 15251, *Indoor Environmental Input Parameters for Design and Assessment of Energy Performance of Buildings Addressing Indoor Air Quality, Thermal Environment, Lighting and Acoustics* (European Committee of Standardization, Brussels, 2007)

EN ISO 7726, *Ergonomics of the Thermal Environment—Instruments for Measuring Physical Quantities* (European Committee of Standardization, Brussels, 1998)

S.A. Klein, W.A. Beckman, J.W. Mitchell et al., *TRNSYS—a Transient System Simulation Program User Manual* (The Solar Energy Laboratory—University of Wisconsin, Madison, 2007)

N. Rossi, *Manuale del termotecnico* (Hoepli, Milano, 2002)

UNI TS 11300-2, *Prestazioni energetiche degli edifici—Parte 2: Determinazione del fabbisogno di energia primaria e dei rendimenti per la climatizzazione invernale e per la produzione di acqua calda sanitaria* (Ente Nazionale Italiano di Unificazione, Milano, 2008)
UNI TS 11300-3, *Prestazioni energetiche degli edifici—Parte 3: Determinazione del fabbisogno di energia primaria e dei rendimenti per la climatizzazione estiva* (Ente Nazionale Italiano di Unificazione, Milano, 2010)

Printed in the United States
By Bookmasters